FUNDAMENTALS OF ELECTROTECHNOLOGY

2 : Practical Guide

By the same author
Fundamentals of Electrotechnology
1: Experimental Approach

FUNDAMENTALS OF ELECTROTECHNOLOGY

2: Practical Guide

H. Wardle
B.Sc., C.Eng., M.I.E.E.

Formerly Deputy Director and Head of the Engineering Department, College of Science and Technology, Kaduna Polytechnic, Nigeria

HUTCHINSON EDUCATIONAL

HUTCHINSON EDUCATIONAL LTD
3 Fitzroy Square, London W1

London Melbourne Sydney Auckland
Wellington Johannesburg Cape Town
and agencies throughout the world

First published 1972

© H. Wardle 1972

This book has been set in cold type by E. W. C. Wilkins and Associates Ltd., printed in Great Britain by litho on smooth wove paper by Anchor Press, and bound by Wm. Brendon, both of Tiptree, Essex

ISBN 0 09 113440 4 (cased)
     0 09 113441 2 (paper)

Contents

Preface

List of Symbols and Abbreviations

1 GENERATION OF SINGLE-PHASE ALTERNATING   1
E.M.F.'s; MAXIMUM, INSTANTANEOUS AND R.M.S.
VALUES; – PHASOR REPRESENTATION

Introduction	1
Generating an alternating e.m.f. in a coil	1
How frequency depends on the number of pole pairs in an alternator and the speed at which it is driven	2
The open-circuit characteristic of an alternator	5
The equation of an alternating wave	6
Examples 1.1	9
Comparing the heating effects of direct and alternating voltages applied across a resistor	10
The effective value of an alternating current waveform	13
Definitions of r.m.s. values of current and voltage	15
Graphical determination of r.m.s. values	15
Mathematical evaluation of the r.m.s. value of a sine wave	16
Determining the average value of a sine wave	18
Form factor	19
Peak factor	20
To determine the r.m.s. values of composite voltage waves	20
Examples 1.2	24
Using phasors to represent alternating quantities	25
What is meant by phase difference?	26
Addition of phasors	27
Subtraction of phasors	29
Combining phasors by resolving into components	30
Phasor diagrams in terms of r.m.s. values	33
Examples 1.3	34

2	RESISTANCE, INDUCTANCE AND CAPACITANCE IN SERIES ALTERNATING CURRENT CIRCUITS	36
	Introduction	36
	Resistance in an alternating current circuit	36
	Examples 2.1	40
	Inductance in an alternating current circuit	40
	Power in a purely inductive circuit	43
	Examples 2.2	45
	Resistance and inductive reactance connected in series	45
	True power, apparent power and power factor	47
	Active and reactive current components	48
	Impedance triangle for R and L in series	48
	Examples 2.3	50
	Capacitance in an alternating current circuit	52
	Power in a purely capacitive circuit	55
	Examples 2.4	57
	Resistance and capacitive reactance connected in series	58
	True power, apparent power and power factor	60
	Impedance triangle for R and C in series	60
	Examples 2.5	61
	Resistance, inductive reactance and capacitive reactance connected in series	62
	The series resonant circuit	64
	Series circuit magnification	67
	Examples 2.6	68
	Alternating current circuit circle diagrams	69
	Examples 2.7	79
3	RESISTANCE, INDUCTANCE AND CAPACITANCE IN PARALLEL ALTERNATING CURRENT CIRCUITS; THE SYMBOLIC NOTATION; ADMITTANCE, CONDUCTANCE AND SUSCEPTANCE	80
	Introduction	80
	Parallel a.c. circuit containing resistance and capacitive reactance	80
	Parallel a.c. circuit containing resistance and inductive reactance	82
	Parallel a.c. circuit containing resistance, inductance and capacitance	83
	Examples 3.1	85
	The parallel resonant circuit	86

Parallel circuit current magnification 88
Examples 3.2 91
Representing phasors by a symbolic notation 92
Examples 3.3 95
Addition of phasors using the j notation 95
Subtraction of phasors using the j notation 96
Examples 3.4 97
Dealing with resistance, inductive reactance and
capacitive reactance in the j notation 98
How may Kirchhoff's second law be applied to a.c. circuits? 100
Examples 3.5 102
What happens if we meet j^2? 103
Multiplying and dividing phasor expressions 105
Power expressed in the j notation 107
Impedances in series 109
Impedances in parallel 110
The branched circuit rule 111
Examples 3.6 112
Admittance, conductance and susceptance 114
Examples 3.7 119

4 THREE PHASE CONNECTIONS, THE ALTERNATOR, 120
AND POWER IN THREE PHASE SYSTEMS

Introduction 120
Generation of three-phase e.m.f.'s 121
How the line and phase voltages in a star-connected 122
alternator vary with increasing field current
Using a phasor diagram to determine the relationship 124
between the line and phase voltages of a star-connected
alternator
Voltage drops and currents in a star-connected three-phase 126
load
Three-phase four-wire distribution system 129
Alternator with a rotating d.c. field system 131
Summation of the induced e.m.f.'s in a mesh-connected 132
alternator
Relationship between the line and phase currents of a 133
mesh-connected alternator supplying a balanced three-
phase load
How the output voltage of an alternator changes with 136
varying conditions of loading

Examples 4.1	140
Measuring power in a three-phase star-connected load, using three wattmeters	141
Measuring power in a three-phase mesh-connected load using three wattmeters	143
Connecting an artificial neutral point	145
The two wattmeter method of measuring three-phase power in a star-connected load	145
Effect of power factor on wattmeter indications when the load is balanced	148
The two wattmeter method of measuring power in a mesh-connected load	150
Examples 4.2	154
Delta to star and star to delta transformations	155
Examples 4.3	158

5 ROTATING FIELDS, THREE-PHASE AND SINGLE-PHASE ALTERNATING CURRENT MOTORS, POWER FACTOR IMPROVEMENT AND ELECTRICITY TARIFFS — 160

Introduction	160
How a rotating field may be produced using a three-phase supply	160
Establishing a four-pole rotating field	162
How an induction motor operates	163
Starting an induction motor	166
Load test on a three-phase induction motor	166
Torque in an induction motor	169
Varying the speed of an induction motor	173
Examples 5.1	174
The synchronous motor	176
Producing a rotating field form a two-phase supply, and the principle of the single-phase induction motor	180
Single phase a.c. series motor	183
Examples 5.2	183
Power factor improvement	184
Electricity tariffs	189
Examples 5.3	191

6 TRANSFORMERS — 193

Introduction	193
The transformer on no load	193

The primary and secondary turns and voltage ratios	194
Deriving the e.m.f. equation for a transformer	195
The transformer phasor diagram on no load	198
No-load losses in a transformer	198
Representing a transformer on no load by an equivalent circuit	199
The transformer as an example of mutual inductance	200
Mutual and self inductances of two coils connected in series	202
Examples 6.1	204
Load test on a single-phase transformer	207
Transformer phasor diagram on load, neglecting winding losses	210
Transformer winding resistances and leakage reactances	211
Testing a transformer under capacitive load conditions	213
Cooling of transformers	215
Examples 6.2	217
Simplified transformer equivalent circuits neglecting iron losses	218
The transfer of resistance, reactance and impedance from one side of a transformer to the other	219
Transformer voltage regulation	222
Open circuit and short circuit transformer tests	224
Examples 6.3	228
Separating transformer core losses	230
Examples 6.4	233
Auto transformers	235
Examples 6.5	238

7 THE SIMPLE POTENTIOMETER AND ELECTRICAL MEASURING INSTRUMENTS — 239

Introduction	239
Connecting and balancing a simple potentiometer	239
Standard cell	241
Comparing the e.m.f.'s of two cells	241
Measurement of current by means of a potentiometer	245
Examples 7.1	246
A commercial form of the direct current potentiometer used to check the calibration of a voltmeter	248
Extending the range of a potentiometer	251
Checking the calibration of an ammeter	254

Measuring low resistance with the aid of a potentiometer	256
Examples 7.2	257
The construction and operation of moving coil instruments	257
Extending the range of a moving coil instrument	260
The moving coil instrument used to measure resistance	264
Examples 7.3	265
The thermoelectric effect	269
Calibrating a thermocouple	269
Using a thermocouple to measure temperature rise	271
Thermocouple ammeter	273
Instrument rectifiers	273
The characteristic curve for a rectifier unit	275
Full wave rectifier for use with a moving coil instrument	276
Examples 7.4	279
Moving iron instruments	279
The use of a current transformer	282
The electrodynamic wattmeter	283
Electrodynamic voltmeters and ammeters	288
Electrostatic voltmeters	289
The fluxmeter	290
Insulation resistance measuring instruments	291
Examples 7.5	293

8 ELECTRONS IN THERMIONIC VALVES — 295

Introduction	295
The diode valve	295
The use of the diode valve as a rectifier	299
Examples 8.1	302
An introduction to the triode valve	303
Triode valve characteristics	304
Thermionic valve parameters	305
How a triode valve may be used as an amplifier	307
The equivalent circuit for a triode amplifier	310
Multi-stage amplifiers	312
The characteristics of a gas filled triode	314
Examples 8.2	318
Introducing the cathode ray oscilloscope	319
How the position of the spot on a cathode ray tube may be varied	321
Shift control	323
Introducing a time base circuit	324
Examples 8.3	327

9	SEMICONDUCTORS AND SEMICONDUCTOR DEVICES	328
	Introduction	328
	How does a semiconductor function?	328
	N-type semiconductors	330
	P-type semiconductors	331
	What happens at the junction between p and n type materials when a voltage is applied?	333
	Stabilising a voltage supply using a semiconductor diode	336
	What is a transistor?	337
	How does a transistor function?	338
	Connecting a transistor with common emitter circuit connections	343
	Examples 9.1	346
10	ILLUMINATION AND ELECTRIC LAMP CIRCUITS	348
	Introduction	348
	How the illumination on a surface depends on the distance of the surface from the light source	348
	Luminous intensity	350
	Units of luminous flux, illumination and luminance	352
	The inverse square and cosine laws of illumination	353
	Examples 10.1	357
	Filament lamps	358
	Electric discharge lamps	359
	Low pressure mercury vapour discharge lamp	359
	High pressure mercury vapour discharge lamp	363
	Low pressure sodium vapour discharge lamp	363
	Light sensitive cells	364
	Examples 10.2	367
11	PRODUCING E.M.F.'s BY CHEMICAL ACTION; ELECTROLYTIC DEPOSITION; SECONDARY CELLS	368
	Introduction	368
	Primary cells	368
	The Leclanché dry cell	370
	Electrolytic deposition	371
	Examples 11.1	374
	Secondary cells	375
	The lead-acid cell	375
	Examples 11.2	378

Alkaline cells 379
Comparison of lead-acid and alkaline cells 380
Examples 11.3 381

Preface

The first book *Fundamentals of Electrotechnology, 1: Experimental Approach* covered basic d.c. circuit theory, magnetism, electrostatics and d.c. machines. This second book, *2: Practical Guide* introduces a.c. circuit theory, single and three-phase networks and machines, measurements, electronics, semiconductors and illumination.

Wherever possible, underlying theories of the topics considered are developed in the International System of Units (SI) through experimental laboratory work. It is intended that the book will supplement students' practical work and be of particular use to those students with restricted opportunities for carrying out experimental work. Examples are included at the end of each section and numerical answers are given.

The range of topics covered in *Fundamental of Electrotechnology 1 and 2* should meet the requirements of students studying electrotechnology in Ordinary National Certificate courses; in courses leading to the intermediate and final examinations in Electrical Installation and Technicians' Work; and also of students attempting Ordinary Technician Diploma examinations as set by the City and Guilds of London Institute for students overseas. While not specifically aimed at students following full-time degree courses, these books ought to be of interest to first year students studying electrotechnology on these courses.

List of symbols and abbreviations

Quantity	Symbol	Unit name	Unit symbol
Acceleration	a or f	metre per second squared	m/s²
Admittance	Y	siemens	S
Amplification factor,			
electronic valve	μ		
common base transistor	α		
common emitter transistor	β		
Angle			
plane	$\alpha, \beta, \ldots$	radian	rad
solid	ω	steradian	sr
Anode slope resistance	r_a	ohm	Ω
Area	a	square metre	m²
Armature conductors	Z		
Capacitance	C	farad	F
Charge or quantity of electricity	Q	coulomb	C
Conductance	G	siemens	S
Conductivity	σ	siemens per metre	S/m
Current,			
steady or r.m.s.	I	ampere	A
instantaneous	i	ampere	A
maximum	I_m	ampere	A
Electric field strength	E	volt per metre	V/m
Electric flux	Q	coulomb	C
Electric flux density	D	coulomb per square metre	C/m²

Quantity	Symbol	Unit name	Unit symbol
Electromotive force,			
steady or r.m.s.	E	volt	V
instantaneous	e	volt	V
maximum	E_m	volt	V
Energy or work	W	joule	J
Frequency	f	hertz	Hz
Illumination	E	lux	lx
Impedance	Z	ohm	Ω
Inductance,			
self	L	henry	H
mutual	M	henry	H
Length	l	metre	m
Luminance	L	candela per square metre	cd/m²
Luminous flux	Φ	lumen	lm
Luminous intensity	I	candela	cd
Magnetic field strength	H	ampere(-turn) per metre	A/m, At/m
Magnetic flux	Φ	weber	Wb
Magnetic flux density	B	tesla	T
Magnetomotive force	F	ampere(-turn)	A, At
Magnification factor	Q		
Mass	m	kilogramme	kg
Mutual conductance	g_m	milliamperes per volt	mA/V
Parallel armature paths	a		
Permeability,			
absolute	μ	henry per metre	H/m
free-space	μ_0	henry per metre	H/m
relative	μ_r		
Permittivity,			
absolute	ϵ	farad per metre	F/m
free-space	ϵ_0	farad per metre	F/m
relative	ϵ_r		
Pole pairs	p		

Quantity	Symbol	Unit name	Unit symbol
Potential difference			
steady or r.m.s.	V	volt	V
instantaneous	v	volt	V
maximum	V_m	volt	V
Power			
true	P	watt	W
apparent	S	voltampere	VA
reactive	Q	reactive voltampere	VAr
Power factor (sine wave)	$\cos\phi$		
Radius	r	metre	m
Reactance			
capacitive	X_c	ohm	Ω
inductive	X_L	ohm	Ω
Resistance	R	ohm	Ω
Resistivity	ρ	ohm metre	Ωm
Slip speed	n_s	revolutions per second	rev/s
Specific heat capacity	c	joules per kilogramme degree Celsius	J/kg°C
Susceptance			
capacitive	B_c	siemens	S
inductive	B_L	siemens	S
Temperature	T, θ	kelvin, degree Celsius	K, °C
Time	t	second	s
Torque	T	newton metre	Nm
Turns	N		
Velocity			
linear	u, v	metres per second	m/s
angular	ω	radian per second	rad/s
rotational	n	revolution per second	rev/s
Wavelength	λ	metre	m

1. Generation of Single-Phase Alternating e.m.f.'s; Maximum, Instantaneous and r.m.s. Values; Phasor Representation

Introduction

In our earlier studies we have considered e.m.f.'s which always act in the same direction. Sources of these e.m.f.'s include batteries and direct current generators. In some applications, to which electricity is put, direct current sources are essential. However, the ease with which an alternating voltage may be changed from one value to another and the simple construction of the cage-type a.c. induction motor are but two reasons why an alternating voltage system may be desirable. In this chapter we begin by considering the generation of alternating e.m.f.'s.

Generating an alternating e.m.f. in a coil

You will recall how when we began the study of generators. In Chapter 8 of Book 1 it was seen that if a single-turn coil was rotated in a magnetic field an e.m.f. was induced in the coil. The e.m.f. acted first in one direction and then in the opposite direction. Indeed, since we required a direct e.m.f. at that stage we had to use a commutator to provide us with the desired direct output from the machine.

In place of the two-part commutator used to obtain a steady output from the machine we may fit two insulated slip rings to the armature shaft and connect the ends of the armature coil to these rings (Fig. 1.1(a)).

As the armature rotates the conductors cut magnetic flux and e.m.f.'s are induced. The slip rings are alternately positive and negative with respect to each other, and the output e.m.f. appearing across the brushes, will be as shown in Fig. 1.1(b). This is termed an alternating e.m.f. and the machine generating it is known as an alternator.

It is usual for an alternator to be driven at such a speed that the alternating e.m.f. goes through one complete set of changes from zero to positive maximum and then to zero, negative maximum and zero again in a short period of time. The number of complete sets of changes which occur per second is termed the frequency of the alternating e.m.f.

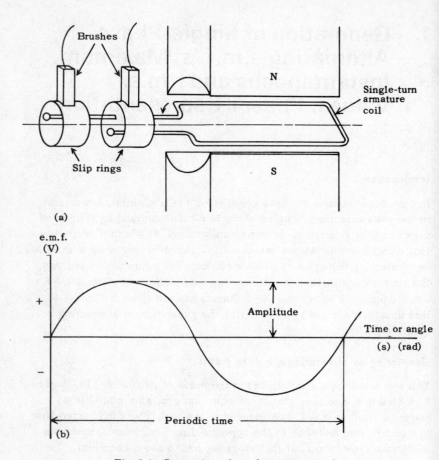

Fig. 1.1. Generation of an alternating e.m.f.

and is measured in hertz (Hz). A supply frequency of 50 Hz thus simply means that 50 complete reversals of the supply current or voltage takes place every second. Most countries have standardised on frequencies of 50 Hz or 60 Hz for power supplies. In radio, television and communications work we often deal with frequencies running into many millions of hertz or MHz.

The time of one complete set or cycle of changes may be referred to as the periodic time of an alternating wave.

How frequency depends on the number of pole pairs in an alternator and the speed at which it is driven

Let us consider an experiment

Experiment 1(i) *To show how the frequency of the e.m.f. induced in an alternator winding depends on the speed at which the machine is driven*

In this experiment a six-pole alternator was used (Fig. 1.2.).

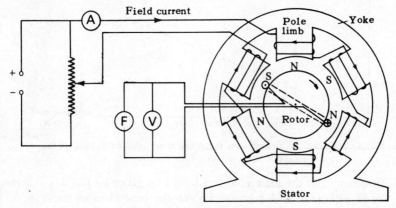

Fig. 1.2. Six-pole alternator.

When dealing with a.c. machines it is often convenient to refer to the rotating part of the machine as the rotor and the stationary part as the stator.

The six field coils on the stator were connected in series across a d.c. supply, using a potentiometer type field current regulator. The alternator rotor carried three coils in which e.m.f.'s were induced but in this present test only one coil was used.

The alternator was driven by a variable speed d.c. shunt connected motor and the speed was increased in steps of 100 rev/min from 700 rev/min to 1500 rev/min. At each step the alternator field current was adjusted until the generated terminal voltage appearing across the rotor winding was 110 volts.

The frequency meter F indicated the frequency of the induced voltage and the results were as tabulated:

Rotor speed (rev/min)	700	800	900	1000	1100	1200	1300	1400	1500
Rotor induced e.m.f. (V)	110	110	110	110	110	110	110	110	110
Frequency (Hz)	35	40	45	50	55	60	65	70	75

It is clear that the frequency of the induced e.m.f. increases in direct proportion to the increase in rotor speed. In a six-pole machine, the e.m.f. induced in a coil undergoes three complete sets of changes in one revolution of the rotor. Thus, if the rotor is driven at n rev/s,

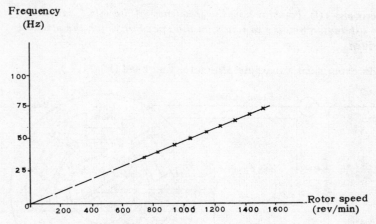

Fig. 1.3. Relationship between the rotor speed and frequency of the induced e.m.f. in a six-pole alternator.

the frequency f of the generated e.m.f. will be given by $f = 3 \times n$ hertz.

For an alternator with p pairs of poles the general expression for frequency is given by $f(\text{hertz}) = p(\text{pole pairs}) \times n(\text{rev/s})$.

Referring to our experimental results we may confirm the validity of this expression. When the speed is 1000 rev/min or 16·67 rev/s, $f = p \times n = 3 \times 16\cdot67 = 50\,\text{Hz}$.

This is the same as the frequency meter indication when the rotor speed was 1000 rev/min.

Example

Calculate from first principles the speed at which a four-pole alternator must be driven to produce a voltage wave of 50 Hz. What will be the periodic time of this wave? Sketch a diagram showing how the alternator induced e.m.f. will vary during one complete revolution of the rotor.

In one revolution of the rotor in a four-pole alternator the induced e.m.f. will pass through two complete sets of changes or cycles (Fig. 1.4.). To produce a frequency of 50 Hz the alternator must be driven at 25 rev/s or 1500 rev/min.

Using the expression, $f = p \times n$

$$n = \frac{f}{p} = \frac{50}{2} = 25\,\text{rev/s} = 1500\,\text{rev/min}$$

The periodic time of the wave $= 1/50\,\text{s} = 0\cdot02$ second.

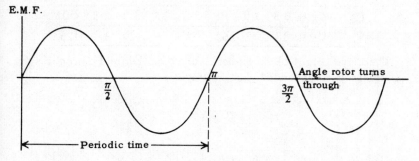

Fig. 1.4. Four-pole alternator-induced e.m.f.

The open-circuit characteristic of an alternator

An alternator open-circuit characteristic depicts the relationship between the field excitation and the resulting terminal voltage on no load. The alternator speed is maintained constant.

Experiment 1(ii) *To plot the open-circuit characteristic of an alternator*

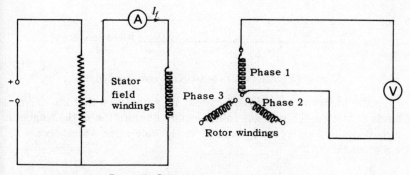

Fig. 1.5. Open-circuit test on an alternator.

The alternator used in this test had three windings on the rotor, with the ends of the windings connected to separate slip rings. With a voltmeter connected across one winding the machine was driven at a constant speed of 1000 rev/min. The field current (I_f) was gradually increased in steps of 0·5 ampere to a maximum of 4 amperes and the rotor phase voltage (V_{ph}) measured at each stage. The observations were as tabulated.

I_f (A)	0	0·5	1·0	1·5	2·0	2·5	3·0	3·5	4·0
V_{ph} (V)	0	18	37	62	77	95	110	118	126

The curve (Fig. 1.6) shows how increasing the field current results in an increased generated phase voltage.

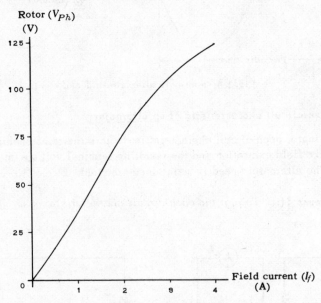

Fig. 1.6. Alternator open-circuit characteristic.

At the higher values of field current it is observed that the curve tends to flatten. This is due to the onset of saturation in the magnetic circuit of the field system — similar to the saturation which occurs in a d.c. machine.

The equation of an alternating wave

Now that you understand the meanings of terms connected with alternating quantities, and the relationship between the number of pole pairs in an alternator, the speed at which it is driven and the resulting frequency of the induced e.m.f.'s, you are ready to have a closer look at alternating waves.

In dealing with alternating current, voltage or other waves where the magnitudes are continually changing as in Fig. 1.7 we will use symbols as follows:

I_m, Φ_m, E_m represent maximum values of current, flux and e.m.f.

i, ϕ, e represent values of current, flux and e.m.f. at any instant, or instantaneous values.

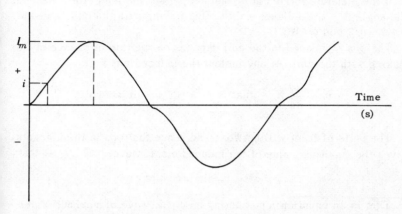

Fig. 1.7. Maximum and instantaneous values of an alternating quantity.

If we consider an armature coil of one turn $A - A'$ rotating in a uniform magnetic field (Fig. 1.8(a)) then the e.m.f. induced at any instant will depend on the rate of change of flux linking with the coil at that instant.

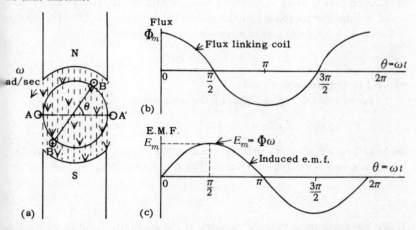

Fig. 1.8. Variation of the magnetic flux linking with a rotating coil and the resulting e.m.f. induced as the armature rotates.

Let Φ_m (Fig. 1.8(b)) be the flux linking with the coil when in the position $A - A'$. Then for any other position of the coil such as $B - B'$ after it has turned through θ radians, the instantaneous value of the flux through the coil is given by

7

$$\phi = \Phi_m \cos \theta \text{ Wb}$$

If the armature rotates at ω radians per second then after t seconds the angle $\theta = \omega t$ radians, and the flux linking with the coil is given by $\phi = \Phi_m \cos \omega t$ Wb.

The e.m.f. induced in the coil depends on the rate of change of flux linking with the coil. At any instant the induced e.m.f.

$$e = -\frac{d\phi}{dt} \text{ V} \quad \text{that is } e = (\Phi_m \omega) \sin \omega t \text{ volts.}$$

The units of $\Phi_m \omega$ will be Wb/s and hence the term in the brackets gives the maximum value of the induced e.m.f. wave.

$$e = E_m \sin \omega t \text{ volts}$$

This is an equation representing an e.m.f. wave of maximum value E_m volts.

Figure 1.8(c) shows the e.m.f. induced by the flux changing as in Fig. 1.8(b). At any instant the magnitude of the induced e.m.f. is given by −(the slope of the flux wave) at that instant.

One complete revolution of the rotor produced one complete cycle of induced e.m.f. Thus if the rotor rotates with an angular velocity of ω rad/s then the frequency of the induced e.m.f., $f = \omega/2\pi$ Hz.

Since $\omega = 2\pi f$ the e.m.f. equation may be written,

$$e = E_m \sin 2\pi f t$$

where f is the frequency in Hz.

If two waves are represented by the equations

$$e_1 = 5 \sin \omega t \quad \text{and} \quad e_2 = 10 \sin 2\omega t$$

then the second wave e_2 will have a maximum value of 10 volts while e_1 has a maximum value of 5 volts. The frequency of e_2 will be double that of e_1.

Note

1. We have assumed that the density of the magnetic field in which the coil rotates (Fig. 1.8(a)) is uniform, and that the change in flux linking with the coil and the induced e.m.f. vary according to cosine and sine curves respectively. In all theory which follows, unless it is specifically stated otherwise, we are dealing with these types of curves.

2. Drawing a sine curve.
    In work which follows you will find it instructive to plot sine

curves and a convenient method of drawing this type of curve is shown in Fig. 1.9.

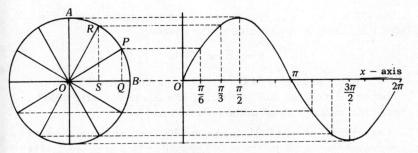

Fig. 1.9. Drawing a sine curve.

The circle has a radius equal to the maximum value of the wave. By striking arcs from A and B round the circumference of the circle, radii at 30° intervals may be inserted.

It will be seen that lengths such as PQ and RS are respectively, $PQ = OP \sin 30°$ and $RS = OP \sin 60°$.

A scale is chosen for the x-axis of the sine curve and instantaneous values projected from the circle at 30° intervals.

Summary

If an alternator having p pairs of poles is driven at n rev/s then the frequency of the induced e.m.f. is given by $f = p \times n$ Hz.

The equation of an alternating e.m.f. wave may be written

$$e = E_m \sin 2\pi f t$$

where E_m is the maximum value and f the frequency in Hz.

Examples 1.1.

1. Describe how an alternating e.m.f. is induced in a coil of insulated wire rotating in a magnetic field.

Explain how slip rings differ from a two-part commutator.

2. How is the frequency of the e.m.f. generated in a single-turn coil rotating in a magnetic field related to the speed of rotation and the number of magnetic poles which produce the field?

At what speed must a coil rotate in a field produced by 6 alternate N and S poles (12 poles in total) if the frequency of the induced e.m.f. is to be 50 Hz? If the coil is rotated at 1200 rev/min what will be the frequency of the e.m.f. wave?

3. Describe how you would carry out a test on an alternator to determine the open-circuit characteristic. Sketch a typical curve and account for the shape.

4. Show that when a coil rotates in a uniform magnetic field the induced e.m.f. may be represented by the equation,

$$e = E_m \sin 2\pi f t.$$

If $e = 60 \sin 628\, t$ what is the frequency of the induced e.m.f. and at what speed must the coil be rotating in a 2-pole magnetic field? What is the maximum value of the induced e.m.f?

5. A square coil of side 0·015 metre has 200 turns and rotates on an axis through the centre points of two opposite sides at 60 rev/s. The axis of rotation is at right angles to a uniform magnetic field of 0·1 tesla.

Plot a curve showing how the e.m.f. induced in the coil varies as the coil rotates. Explain with reasons why the torque required to rotate the coil would be greater if the ends of the coil were connected across a low resistor than when they are open circuited.

6. A rectangular coil of 50 turns is 0·25 m × 0·5 m.

It rotates at 1000 rev/min (about an axis passing through the centre points of the two shorter sides) in a uniform magnetic field of 0·3 tesla. Write down an equation representing the induced e.m.f. when the plane of the coil makes an angle of (a) 30° with the direction of the field and (b) 90° with the direction of the field.

7. Write down the equation of an alternating current wave which has a maximum value of 6 amperes and a frequency of 50 Hz.

Comparing the heating effects of direct and alternating voltages applied across a resistor

You already know how to carry out calculations concerned with the heating effect of a direct current. What about the heating effect of an alternating current?

Let us suppose that we apply an alternating voltage wave as represented by the curve in Fig. 1.10. across a resistor immersed in a calorimeter of water. As current flows the temperature of the water may be expected to rise.

What value of direct voltage connected across the resistor would produce the same amount of heat as the alternating voltage in the same time?

Obviously the steady direct voltage would be less than V_m since

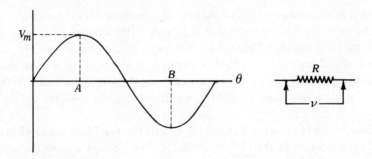

Fig. 1.10. Alternating voltage applied across a resistor.

the maximum value of the alternating voltage wave only acts as the the instants corresponding to *A* and *B* in the diagram. At all other instants the applied voltage is less than this.

Let us attempt to solve our problem by carrying out an actual experiment.

Experiment 1(iii) *To determine the value of direct voltage which when applied across a resistor will produce the same heating effect as a given alternating voltage in the same time*

A circuit was connected as in Fig. 1.11.

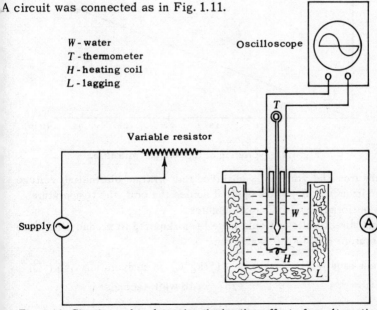

Fig. 1.11. Circuit used to determine the heating effect of an alternating voltage applied across a resistor.

11

In this experiment the heating coil of a calorimeter containing 0·1 kg of water was connected to a 12 volt 50 Hz supply.

A cathode ray oscilloscope (C.R.O.) was connected across the terminals of the coil. The C.R.O. depicts on a screen the actual waveform of the voltage drop across the coil and the maximum value of the wave may be measured by placing a calibrated scale over the oscilloscreen.

The variable resistor was adjusted until the maximum value of the voltage drop across the coil was 4 volts. The voltage across the coil was maintained at this value throughout the test and observations of the rise in the water temperature and time were recorded.

Resistance of calorimeter coil: 1·34 ohms

Time (min)	0	5	10	15	20	25	30
Temperature rise (°C)	0	4	8	12·5	16	20	23

The results are plotted in Fig. 1.12.

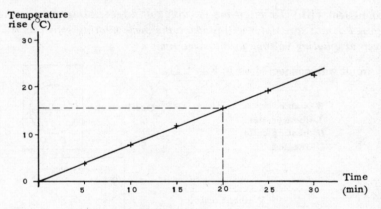

Fig. 1.12. Variation of temperature with time.

Now from the graph it is observed that with an alternating voltage of maximum value 4 volts applied across the coil, the temperature rises uniformly by 16°C in 20 minutes.

What direct voltage would have been required to produce the same temperature rise in the same time?

Heat required = m(kg) × c(J/kg°C) × temperature rise (°C)
= 0·1 × 4187 × 16 watt seconds

Power = $\dfrac{\text{watt seconds}}{\text{seconds}}$

$$\text{Power} = \frac{0.1 \times 4187 \times 16}{20 \times 60} \text{ watts}$$

$$= 5.6 \text{ watts}$$

Now power,

$$P = \frac{V^2}{R} \therefore V = \sqrt{R \times P}$$

$$= \sqrt{1.34 \times 5.6} = \sqrt{7.5}$$

$$= 2.74 \text{ volts.}$$

It is thus clear from our experimental results that an alternating voltage wave of maximum value 4 volts applied across a resistor produces the same heating effect as a steady voltage of 2·74 volts (Fig. 1.13.) applied across the same resistor.

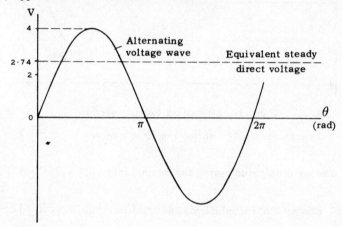

Fig. 1.13. Experimental comparison of alternating and direct voltages.

The effective value of an alternating current waveform

The effectiveness of an alternating current waveform is measured by its heating effect. We will use the symbol I to represent this effective value.

Consider an alternating current with a waveform as shown in Fig. 1.14 flowing through a resistor of R ohms.

Let us divide the wave into n equal intervals with $i_1, i_2 \ldots i_n$ as the mid-ordinates of these intervals.

Now during each interval, energy is dissipated in the form of heat.

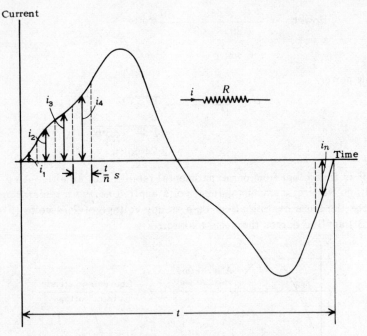

Fig. 1.14. Alternating current waveform.

Energy dissipated during the first interval $= i_1^2 R \dfrac{t}{n}$ J

Energy dissipated during the second interval $= i_2^2 R \dfrac{t}{n}$ J

Energy dissipated during the third interval $= i_3^2 R \dfrac{t}{n}$ J

Total energy dissipated in t seconds

$$= i_1^2 R \dfrac{t}{n} + i_2^2 R \dfrac{t}{n} + i_3^2 R \dfrac{t}{n} + \ldots i_n^2 R \dfrac{t}{n} \text{ J}$$

$$= R t \dfrac{(i_1^2 + i_2^2 + i_3^2 + \ldots i_n^2)}{n} \text{J}$$

Let I represent the direct current which would produce the same heating effect as the alternating current.

In t seconds, heat energy produced $= I^2 R t$ watt seconds or joules.

Since the same heat energy is to be produced by the direct and alternating current waves,

$$I^2Rt = \frac{Rt(i_1^2 + i_2^2 + i_3^2 + \ldots i_n^2)}{n}$$

$$\therefore I = \sqrt{\frac{(i_1^2 + i_2^2 + i_3^2 + \ldots i_n^2)}{n}}$$

= square root of the mean of the squares of successive ordinates

= root—mean—square value

I = r.m.s. value

The effective value of an alternating current is thus often referred to as the r.m.s. value.

Similarly the effective value of an alternating voltage wave may be determined graphically using the expression

$$V = \sqrt{\frac{(v_1^2 + v_2^2 + v_3^2 + \ldots v_n^2)}{n}}$$

Definitions of r.m.s. values of current and voltage

The effective, or r.m.s., value of an alternating current wave is that steady value of direct current which, when flowing through a given resistor, produces the same heating effect as the alternating current under the same conditions.

The effective or r.m.s. value of an alternating voltage wave is that steady value of direct voltage which, when applied across a given resistor, produces the same heating effect as the alternating voltage under the same conditions.

Graphical determination of r.m.s. values

Let us determine graphically the effective, r.m.s. or equivalent direct voltage to an alternating sinusoidal voltage wave of maximum value 4 volts.

Figure 1.15 shows the voltage wave and also a curve of v^2 obtained by taking equally spaced ordinates at v and squaring these.

The effective or r.m.s. value,

$$V = \sqrt{\frac{(v_1^2 + v_2^2 + v_3^2 + \ldots v_n^2)}{n}}$$

$$= \sqrt{\frac{47 \cdot 4}{6}} = 2 \cdot 81 \text{ volts}$$

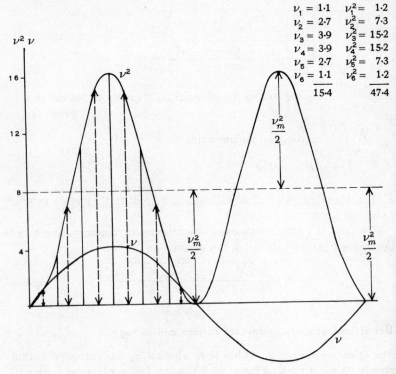

Fig. 1.15. Determining the r.m.s. value of a wave.

This value is seen to agree quite closely with the value of 2·74 volts as determined in **Experiment 1(iii)** using a calorimeter.

Mathematical evaluation of the r.m.s. value of a sine wave

Now let us see how easily we may determine mathematically the r.m.s. value of a sine wave.

Let
$$\nu = V_m \sin \theta$$

$$\therefore \nu = V_m^2 \sin^2 \theta$$

$[2 \sin \theta = 1 - \cos 2\theta]\quad = V_m^2 \dfrac{(1 - \cos 2\theta)}{2}$

$$= \frac{V_m^2}{2} - \frac{V_m^2 \cos 2\theta}{2} \ldots \ldots (1)$$

This equation may be identified as the curve of v^2 in Fig. 1.15. It has a steady component

$$\frac{V_m^2}{2} = \frac{16}{2} = 8 \text{ volts},$$

which is represented by the broken line, and also an alternating component $\frac{V_m^2 \cos 2\theta}{2}$ of maximum value

$$\frac{V_m^2}{2} = \frac{16}{2} = 8 \text{ volts}.$$

The alternating component has as its axis the broken line and it has a frequency twice that of the curve representing v.

Now,

$$V = \sqrt{\frac{v_1^2 + v_2^2 + v_3^2 + \ldots v_n^2}{n}}$$

$$= \sqrt{\text{average ordinate of the } v^2 \text{ curve}}$$

In equation (1) above, the average value of the first term is

$$\frac{V_m^2}{2}$$

and the average value of the second term

$$\frac{V_m^2 \cos 2\theta}{2} = 0.$$

Hence the r.m.s. value,

$$V = \sqrt{\frac{V_m^2}{2} - 0}$$

$$= \frac{V_m}{\sqrt{2}}$$

When dealing with either current or voltage sine waves the r.m.s. value is given by,

$$\text{r.m.s. value} = \frac{\text{maximum}}{\sqrt{2}}$$

Thus the r.m.s. value V of a sine wave having a maximum value of 4 volts,

$$V = \frac{V_m}{\sqrt{2}} = \frac{4}{\sqrt{2}} = 2\cdot 83 \text{ volts.}$$

This is the correct value. The two values determined from the experiment using a calorimeter and also from the graphical solution are both slightly low, the calorimeter error being caused by heat losses and in the graphical method errors in taking measurements occur. Electrical laboratory and switchboard instruments are normally calibrated to give indications in r.m.s. values.

Determining the average value of a sine wave

Referring to Fig. 1.15 we see that the average value of the wave during one half cycle is given by,

$$\frac{\nu_1 + \nu_2 + \nu_3 + \ldots \nu_6}{6} = \frac{15\cdot 4}{6} = 2\cdot 6\text{ V}$$

Could we calculate the average value mathematically?
Let
$$\nu = V_m \sin \theta.$$

Consider one half of a sine curve (Fig. 1.16).

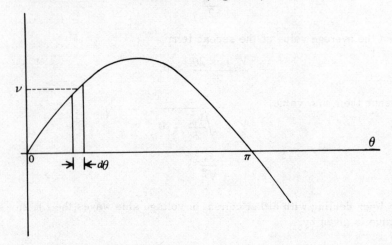

Fig. 1.16. Determining the average value of a sine wave.

$$\text{Average value} = \frac{\text{area}}{\text{base}}$$

$$\text{area} = \int_0^\pi \nu\, d\theta = V_m \int_0^\pi \sin\theta\, d\theta$$

$$= -V_m [\cos\theta]_0^\pi = -V_m [\cos\pi - \cos 0]$$

$$= 2V_m$$

$$\text{Average value} = \frac{\text{area}}{\text{base}} = \frac{2V_m}{\pi}$$

$$= 0\cdot 636 \times \text{maximum value}$$

Thus for a sine wave of maximum value 4 volts,

$$\text{the average value} = \frac{2 \times 4}{\pi} = 2\cdot 55 \text{ volts.}$$

This agrees closely with the graphically determined value of 2·6 volts.

Form factor

It will be seen that the r.m.s. value is higher than the average value (Fig. 1.17). The ratio of the r.m.s. value to the average value of any wave is known as the form factor of the wave.

For a sine wave,

$$\text{form factor} = \frac{\text{r.m.s.}}{\text{average}} = \frac{\text{maximum}}{\sqrt{2}} \times \frac{\pi}{2 \times \text{maximum}}$$

$$= 1\cdot 11$$

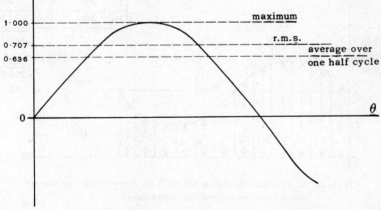

Fig. 1.17. Maximum, r.m.s. and average values.

Peak factor

The peak factor is defined as the ratio of the maximum value to the r.m.s. value.

$$\text{For a sine wave the peak factor} = \frac{\text{maximum}}{\text{r.m.s.}} = \frac{\text{maximum} \sqrt{2}}{\text{maximum}} = 1{\cdot}41$$

A knowledge of the peak factor is necessary when dealing with the maximum stress to which insulation in a.c. circuits may be subjected.

To determine the r.m.s. values of composite voltage waves

In dealing with this topic we will consider two composite waves.

(a) *Alternating voltage wave superimposed on a steady direct voltage wave*

Let us consider a battery and alternating voltage source connected in series across a resistor as in Fig. 1.18(a).

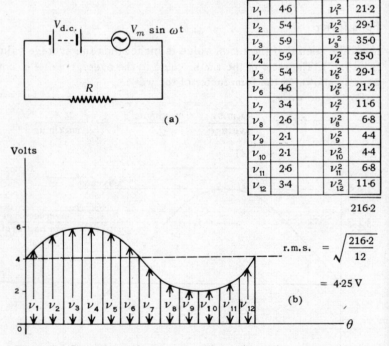

v_1	4·6		v_1^2	21·2
v_2	5·4		v_2^2	29·1
v_3	5·9		v_3^2	35·0
v_4	5·9		v_4^2	35·0
v_5	5·4		v_5^2	29·1
v_6	4·6		v_6^2	21·2
v_7	3·4		v_7^2	11·6
v_8	2·6		v_8^2	6·8
v_9	2·1		v_9^2	4·4
v_{10}	2·1		v_{10}^2	4·4
v_{11}	2·6		v_{11}^2	6·8
v_{12}	3·4		v_{12}^2	11·6
				216·2

$$\text{r.m.s.} = \sqrt{\frac{216{\cdot}2}{12}} = 4{\cdot}25 \text{ V}$$

Fig. 1.18. Composite voltage wave with an alternating component superimposed on a direct component.

If the direct component has a steady value of 4 V and the alternating component has a maximum value of 2 V then by drawing a scale diagram as in Fig. 1.18(b) the r.m.s. value of the wave may be evaluated. By dividing the wave into twelve intervals and squaring the mid-ordinate of each interval the r.m.s. value is determined as being 4·25 volts.

Could we calculate this value directly without drawing a scale diagram?

If V is the r.m.s. value of the composite wave then when this voltage is applied across R the rate at which heat is produced, or the power, will be given by V^2/R.

The equation of the composite applied voltage wave is

$$v = V_{d.c.} + V_m \sin \omega t$$

$$\therefore \frac{V^2}{R} = \frac{(V_{d.c.} + V_m \sin \omega t)^2}{R}$$

that is

$$V^2 = V_{d.c.}^2 + V_m^2 \sin^2 \omega t + 2 V_{d.c.} V_m \sin \omega t$$

$$= V_{d.c.}^2 + \frac{V_m^2}{2} - \frac{V_m^2}{2} \cos 2\omega t + 2 V_{d.c.} V_m \sin \omega t$$

The average values of the two latter terms are zero and hence,

$$V^2 = V_{d.c.}^2 + \frac{V_m^2}{2}$$

$$= V_{d.c.}^2 + V_{r.m.s.}^2$$

or $$V = \sqrt{V_{d.c.}^2 + V_{r.m.s.}^2}$$

When appropriate values for the direct and alternating current components are inserted we have

$$V = \sqrt{(4)^2 + \left(\frac{2}{\sqrt{2}}\right)^2}$$

$$= \sqrt{16 + 2} = \sqrt{18} = 4 \cdot 25 \text{ volts.}$$

This is in agreement with the value obtained using the graphical method.

Students are urged to repeat the graphical and theoretical methods in order to obtain the r.m.s. value of the composite wave when the alternating component is increased to have a maximum value of 4 volts.

(b) *Two alternating waves of the same frequency and in phase with one another.*

Consider two alternating voltages connected in series across a resistor as in Fig. 1.19(a).

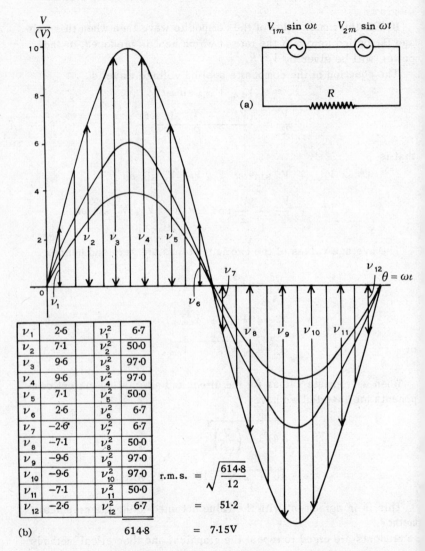

Fig. 1.19. Composite voltage wave comprising two sinusoidal waves superimposed.

Two alternating waves are in phase when they both increase through zero at the same instant.

Let us assume that the component waves have maximum values of 6 V and 4 V respectively. By drawing to scale a diagram as in Fig. 1.19(b) the r.m.s. value of the combined wave may be evaluated. Dividing the combined wave into twelve intervals and squaring the mid-ordinate of each interval the r.m.s. value is found to be 7·15 volts.

How could we calculate this value directly without drawing a scale diagram?

If the r.m.s. value of the resultant wave is denoted by V then when this voltage is applied across R the rate at which heat is produced or the power, will be given by V^2/R.

The equation of the composite voltage wave is

$$V_{1m} \sin \omega t + V_{2m} \sin \omega t = (V_{1m} + V_{2m}) \sin \omega t$$

$$\therefore \frac{V^2}{R} = \frac{(V_{1m} + V_{2m})^2 \sin^2 \omega t}{R}$$

$[2\sin^2\theta = 1 - \cos 2\theta]$

$$\therefore V^2 = \frac{(V_{1m} + V_{2m})^2}{2}(1 - \cos 2\omega t)$$

$$= \frac{(V_{1m} + V_{2m})^2}{2} - \frac{(V_{1m} + V_{2m})^2}{2} \cos 2\omega t$$

The mean value of the later term will be zero.

$$\therefore V = \frac{V_{1m} + V_{2m}}{\sqrt{2}}$$

Inserting the given values for V_{1m} and V_{2m} we have

$$V = \frac{6+4}{\sqrt{2}} = \frac{10}{\sqrt{2}} = 7\cdot 07 \text{ V}$$

This is in close agreement with the graphical value already obtained.

You are urged to repeat both the graphical and calculation methods to obtain the r.m.s. value of the composite wave when the component waves have maximum values of 3 volts and 5 volts respectively.

Summary

The average value of an alternating current $= \dfrac{i_1 + i_2 + i_3 + \ldots i_n}{n}$

The r.m.s. value of an alternating current $= \sqrt{\dfrac{i_1^2 + i_2^2 + i_3^2 + \ldots i_n^2}{n}}$

The r.m.s. value of an alternating current is the value of the direct current which produces the same heating effect as the alternating current.

Form factor $= \dfrac{\text{r.m.s.}}{\text{average}}$. Peak factor $= \dfrac{\text{maximum}}{\text{r.m.s.}}$.

For sine waves:

$\text{r.m.s.} = \dfrac{\text{maximum}}{\sqrt{2}}$; average $= 0{\cdot}636 \times$ maximum

form factor $= \dfrac{\text{r.m.s.}}{\text{average}} = 1{\cdot}11$

peak factor $= \dfrac{\text{maximum}}{\text{r.m.s.}} = 1{\cdot}41$

Average and r.m.s. values of other alternating quantities may be determined using expressions similar to those for alternating currents.

Examples 1.2

1. Describe in your own words what is meant by the effective value of (a) an alternating current wave and (b) an alternating voltage wave.

2. Why is the r.m.s. value of an alternating current of importance?

A stepped a.c. wave varies as follows over the positive half cycle. The negative half cycle is similar.

Current (A)	5	10	5	5
Time interval (s)	0·0 to 0·005	0·005 to 0·01	0·01 to 0·015	0·015 to 0·02

Calculate the form factor for the wave and also the frequency. Which would produce the greater heating effect when passing through a given resistor; the stepped wave or a sinusoidal alternating current wave having a maximum value of 8 A and a frequency of 50 Hz?

3. What do you understand by the terms, frequency, periodic time and

hertz when applied to an alternating current waveform? If a sinusoidal a.c. waveform has a maximum value of 10 amperes determine graphically its r.m.s. value.

4. Write down expressions for alternating current and voltage waves with frequencies of 50 Hz which would produce the same heating effects with a given resistor as direct voltages and currents of 60 volts and 10 amperes respectively.

5. Why is the root—mean—square value of an alternating current the value commonly used?

Three 2-volt cells and a 50 Hz sinusoidal alternating voltage of maximum value 4 volts are connected in series. Plot the resulting voltage wave appearing across the complete circuit over one cycle of the alternating component and determine the ratio of the r.m.s. to the mean value of the combined wave.

6. If an alternating current is represented by the equation, $100 \sin 314t$ where the maximum current is 100 amperes, calculate, (a) the r.m.s. value of the current, (b) the frequency and (c) the value of the current 0·001 second after increasing through zero.

7. What is meant by the peak factor of an alternating wave? A triangular wave of frequency 50 Hz has a maximum value of 100 volts. Determine graphically or otherwise the form factor and peak factor.

Using phasors to represent alternating quantities

In mechanical engineering science we find it convenient to use vectors to represent forces. A vector is a line with an arrow head at one end pointing in the direction in which the force acts and the length of the vector represents the magnitude of the force.

In alternating current work we use rotating vectors which we term phasors to represent alternating quantities. At this stage in our studies the length of the phasor is chosen to represent the maximum value of the alternating quantity being considered. The phasor rotates in an anticlockwise direction and the speed of rotation in rev/s is equal to the supply frequency in Hz. A solid arrow head will be used on phasors which represent currents and an open arrow head on phasors which represent voltages.

In Fig. 1.20 a phasor $O - A$ rotates about O. It is clear from the diagram that at any instant $OP = OA \sin \theta$.

If the length OA is chosen to represent the maximum value of an alternating voltage wave, then at any instant the value $v = V_m \sin \theta$ will be represented by OP which is the projection of the phasor OA on the vertical axis at that instant.

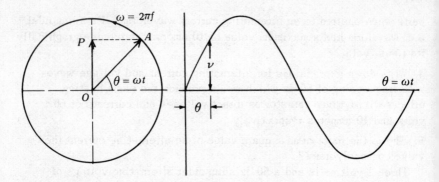

Fig. 1.20. Representing an alternating quantity by a rotating vector or phasor

What is meant by phase difference?

Suppose that we have two alternating current waveforms as expressed by the equations,

$$i_1 = 6 \sin \theta$$
$$i_2 = 8 \sin \left(\theta + \frac{\pi}{3}\right)$$

What does this mean?

The figures 6 and 8 indicate that the maximum values of the two currents are 6 and 8 respectively.

It is usual to draw phasor diagrams for the instant when $t = 0$, which makes $\theta = \omega t$ in these two equations zero.

Let us draw the phasors and waves representing i_1 and i_2.

(a) $\qquad i_1 = 6 \sin \theta$

At the instant when $t = 0$, $\sin \theta = \sin \omega t = \sin 0 = 0$ hence

$$i_1 = 0$$

The current i_1 is passing through a zero value and the projection of the phasor on the vertical axis must be zero. The phasor representing the current is horizontal and shown by OA in Fig. 1.21(a).

(b) $\qquad i_2 = 8 \sin \left(\theta + \frac{\pi}{3}\right)$

At the instant when $t = 0$,

$$\sin \left(\theta + \frac{\pi}{3}\right) = \sin \left(\omega t + \frac{\pi}{3}\right) = \sin \left(\frac{\pi}{3}\right)$$

$$= \frac{\sqrt{3}}{2} = 0\cdot 866$$

$$\therefore i = 8 \times 0\cdot 866 = 6\cdot 928 \text{ amperes.}$$

The projection of the phasor on the vertical axis to represent this current must be 6·928 amperes. The phasor is represented by OB in Fig. 1.21(a).

The two current waves are shown in Fig. 1.21(b).

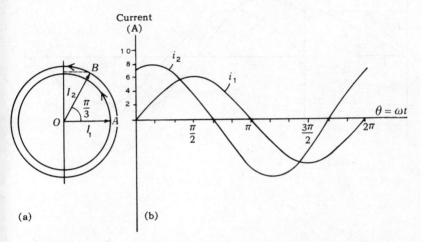

Fig. 1.21. Two waves with a phase displacement of $\pi/3$ radians.

It will simplify our work now and later if we denote a current phasor such as OB by the symbol **I**. This bold italic **I** stands for the words 'the current phasor'. Other phasor quantities will be denoted as the need arises by using the appropriate bold italic symbol.

When considering two or more waves, the wave which first reaches a positive maximum value is said to be the leading wave.

It is seen that the wave represented by I_2 reaches its maximum positive value before the wave represented by I_1 and hence I_2 leads I_1.

Addition of phasors

Let us consider two coils connected in series with e.m.f.'s induced as in Fig. 1.22(a).

If the equations $e_2 = 8\sin(\theta + \pi/4)$ and $e_1 = 6\sin\theta$ represent the two e.m.f.'s what will be the total e.m.f. induced and acting within the coils from A to B?

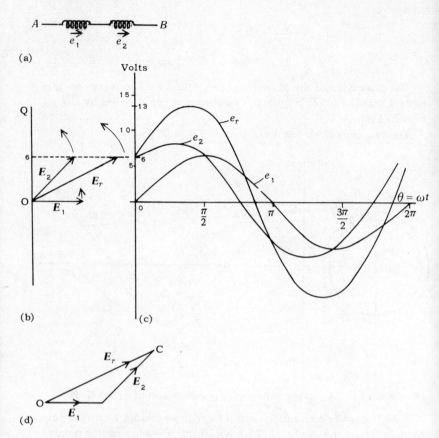

Fig. 1.22. Addition of two phasor quantities.

E_1 and E_2 are the phasors representing the e.m.f.'s. They are drawn in Fig. 1.22(b) at the instant when t and hence θ are zero. Both waves are shown plotted in Fig. 1.22(c) and the resultant wave e_r is obtained by combining the ordinates of the component waves e_1 and e_2. At any instant $e_r = e_1 + e_2$.

It is seen that the resultant wave is also a sine wave with a maximum value of 13 volts and an instantaneous value of 6 volts when t is zero. The phasor representing the resultant wave must have a length representing 13 volts and a projection on the vertical axis $O-Q$ of 6 volts when t is zero. E_r as drawn in Fig. 1.22(b) fulfills these requirements.

It is seen that the phasor E_r representing the resultant wave could have been obtained by adding the individual phasors E_1 and E_2 graphically as in Fig. 1.22(d).

This leads us to a general rule for the addition of phasors.

Draw the tail of one phasor (in this case E_2) to the head of another (in this case E_r) and join the two extreme points as O and c to give the resultant phasor E_r.

Subtraction of phasors

Now let us consider another example where two coils connected in series have e.m.f.'s induced as in Fig. 1.23(a).

$$e_1 = 10 \sin \omega t \qquad e_2 = 8 \sin(\omega t - \pi/3)$$

We are required to determine the total e.m.f. induced and acting within the coils from A to B.

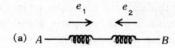

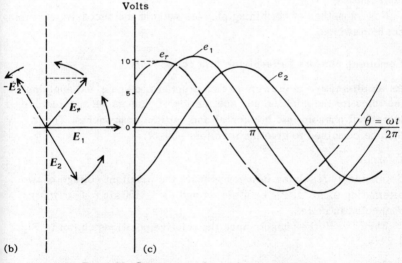

Fig. 1.23. Subtraction of two phasor quantities.

Considering the problem graphically E_1 and E_2 in Fig. 1.23(b) are phasors representing the e.m.f.'s. In Fig. 1.23(c) the two individual waves are plotted. At any instant the total e.m.f. acting from A to B is the value of e_1 at that instant less the value of e_2 at that instant. By subtracting the ordinates of e_2 from e_1 we may construct the resultant wave e_r. This resultant wave is seen to be a sine wave with a maximum value of 9·5 volts and at the instant when t is zero it has a value of 7 volts. These two values enable us to draw the phasor E_r of length representing 9·5 volts and with a projection on the vertical axis of 7 volts.

Now from this phasor diagram we can devise a rule for subtracting one phasor from another. The rule is to reverse the direction of the phasor being subtracted and proceed as in addition.

It will be seen that if E_2 is reversed to become $-E_2$ then E_r is the resultant of $E_1 + (-E_2)$.

$$E_r = E_1 + (-E_2)$$

We are now able to determine the resultant magnitude of a number of phasors by simply drawing phasor diagrams to scale and applying the rules for the addition and subtraction of phasors. It is no longer necessary for us to draw complete wave diagrams. Do not forget however that phasors represent waves. On certain occasions when dealing with phasors you may find it helpful to visualise the waves they represent.

Now, a method of combining phasors without the necessity of making scale drawings.

Combining phasors by resolving into components

As an alternative to drawing phasor diagrams to scale and applying the rules for the addition and subtraction of phasors we may resolve individual phasors into horizontal and vertical components. These may then be combined to give the resultant phasor.

Example

Determine by resolving into components the resultant voltage of two alternating e.m.f.'s, $e_1 = 60 \sin \omega t$ and $e_2 = 100 \sin (\omega t + \pi/4)$ connected in series.

When $t = 0$ the phasors have the relative positions shown in Fig. 1.24(a).

Horizontal component of $E_1 = 60$ V

Horizontal component of $E_2 = 100 \cos \pi/4 = 100/\sqrt{2} = 70\cdot7$ V

Total horizontal components of the resultant phasor $= 60 + 70\cdot7 = 130\cdot7$ V.

Vertical component of $E_1 = 0$

Vertical component of $E_2 = 100 \sin \pi/4 = 100/\sqrt{2} = 70\cdot7$ V

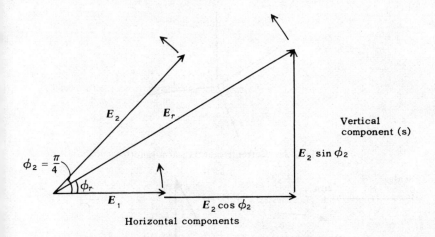

Fig. 1.24. Resolving two phasors into components.

Total vertical component of the resultant phasor = $0 + 70.7 = 70.7$ V.

$$\text{Resultant phasor magnitude} = \sqrt{(\text{horizontal components})^2 + (\text{vertical components})^2}$$

$$= \sqrt{130.7^2 + 70.7^2}$$

$$= 149 \text{ V}$$

Angle between E_r and E_1.

$$\phi_r = \tan^{-1} \frac{\text{vertical components}}{\text{horizontal components}}$$

$$= \tan^{-1} \frac{70.7}{130.7} = \tan^{-1} 0.54$$

$$= 0.5 \text{ rad}$$

$$\therefore e_r = 149 \sin(\omega t + 0.5)$$

Example

Three currents,

$$i_1 = 40 \sin \omega t$$
$$i_2 = 60 \sin(\omega t + \pi/6)$$
$$i_3 = 80 \sin(\omega t + 2\pi/3)$$

meet at a junction as shown in Fig. 1.25. Determine the resultant current

$$i_4 = i_1 + i_2 + i_3$$

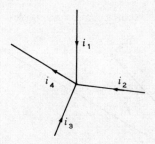

Fig. 1.25. Currents meeting at a junction.

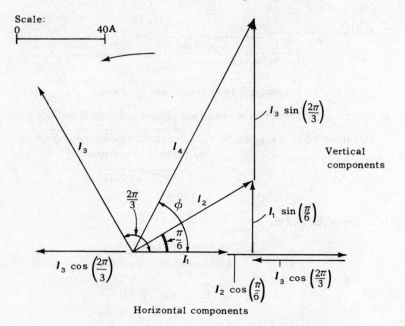

Fig. 1.26. Resolving phasors into components.

When $t = 0$ we have the condition shown in Fig. 1.26.

Horizontal components of current phasors,

$$I_1 = 40 \text{ A}$$
$$I_2 = 60 \cos \pi/6 = 60 \times \frac{\sqrt{3}}{2} = 52 \text{ A}$$
$$I_3 = 80 \cos 2\pi/3 = -80 \times \frac{1}{2} = -40 \text{ A}$$

Horizontal component of resultant, $I_4 = 40 + 52 - 40 = 52$ A.

Vertical components of current phasors,

$$I_1 = 0 \text{ A}$$

$$I_2 = 60 \cos \pi/6 = 60 \times \frac{1}{2} = 30 \cdot 0 \text{ A}$$

$$I_3 = 80 \sin 2\pi/3 = 80 \times \frac{\sqrt{3}}{2} = 69 \cdot 4 \text{ A}$$

Vertical component of resultant, $I_4 = 0 + 30 + 69 \cdot 4 = 99 \cdot 4$ A.
Maximum value of resultant current $= \sqrt{52^2 + 99 \cdot 4^2} = 112$ A.
Angle between I_4 and I_1,

$$\phi = \tan^{-1} \frac{\text{vertical components}}{\text{horizontal components}}$$

$$= \tan^{-1} \frac{99 \cdot 4}{52} = \tan^{-1} 1 \cdot 91$$

hence,
$$= 1 \cdot 1 \text{ radians}$$

$$i_4 = 112 \sin (\omega t + 1 \cdot 1) \text{ amperes}$$

Phasor diagrams in terms of r.m.s. values

Whenever we represent alternating quantities by means of phasors we must remember that these phasors represent waves and it often helps to visualise the waves which they represent.

In drawing phasor diagrams we have chosen the length of a phasor in accordance with the maximum value of the alternating quantity which it represents.

In alternating current theory we are generally interested in r.m.s. values and it is quite in order and indeed usual to draw phasor diagrams with the lengths of the phasors representing r.m.s. values. We have exactly the same form of diagram only the scales are different.

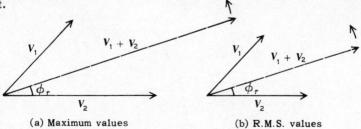

(a) Maximum values   (b) R.M.S. values
Fig. 1.27. Phasor diagrams in maximum and r.m.s. values.

Figure 1.27 shows two diagrams both of which convey the same information, the lengths of the phasors in (b) being equal to those in (a) divided by $\sqrt{2}$.

In future work when we refer to an alternating quantity as being 60 volts or 3 amperes we will assume that these are r.m.s. values.

Summary

A rotating phasor may be used to represent an alternating quantity. The speed of rotation in rev/s corresponds to the frequency of the alternating quantity in Hz, and the instantaneous value is given by the projection of the phasor on the vertical axis. The length of a phasor represents the maximum value of the alternating quantity.

Phasors may be combined by (a) adding or subtracting graphically or (b) resolving into horizontal and vertical components.

Phasor diagrams may be drawn in terms of r.m.s. or maximum values.

Examples 1.3.

1. Describe clearly how an alternating quantity may be represented by a rotating phasor. How is the frequency of the quantity indicated? Draw the position of a phasor depicting the instantaneous value of a sinusoidal e.m.f. having a maximum value 25 volts, at the instant 0·033 second after it is increasing positively through zero. The frequency of the e.m.f. is 50 Hz.

2. What is meant when two e.m.f.'s are referred to as being $\pi/4$ radians out of phase with each other? Represent by phasors the two waves $v_1 = V_m \sin \omega t$ and $v_2 = V_m \sin(\omega t + \pi/2)$. If V_m is 50 volts sketch the waves and indicate clearly which wave is leading.

3. Three currents represented by the equations:

$$i_1 = 8 \sin \theta$$
$$i_2 = 12 \sin(\theta + \pi/2)$$
$$i_3 = 10 \sin(\theta - \pi/3)$$

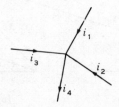

Fig. 1.28. Example 3.

flow in conductors meeting at a junction as in Fig. 1.28. Determine by graphical means and also by resolving into components the current i_4. Express this current in the form of an equation. Determine the phase difference between i_1 and i_4. What will be the time delay between i_1 and i_4 reaching their maximum positive values if the frequency is 50 Hz?

4. Three sinusoidal e.m.f.'s are induced in coils connected in series as shown in Fig. 1.29.

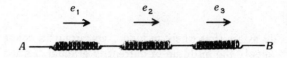

Fig. 1.29. Example 4.

If $e_1 = 30 \sin \omega t$, $e_2 = 50(\sin \omega t + \pi/4)$ and $e_3 = 30 \sin(\omega t - \pi/3)$ determine the resultant e.m.f. acting within the coils from A to B.

5. Give a rule by which one phasor may be subtracted from another. If in the previous example the e.m.f.'s were induced in the coils as in Fig. 1.30. determine the total e.m.f. acting from A to B and express this e.m.f. in the form of an equation.

Fig. 1.30. Example 5.

6. Three sinusoidal alternating voltages are as follows:

$$v_1 = 30 \sin \omega t, \quad v_2 = 50 \sin(\omega t + \pi/4),$$
$$v_3 = 60 \sin(\omega t - \pi/3).$$

Determine by calculation the sum of the three voltages giving your answer in the form $v = V \sin(\omega t \pm \phi)$. Verify your answer by means of a phasor diagram drawn to scale.

7. If the equations in example 6 include the maximum values of the voltage waves, draw the phasor diagram to scale in terms of r.m.s. values.

2. Resistance, Inductance and Capacitance in Series Alternating Current Circuits

Introduction

You are already aware that when an inductive circuit is connected across a d.c. supply the build up of current is gradual. As the current grows the change is opposed by a self-induced back e.m.f. Now, if an inductor is connected into a circuit which includes a source of alternating e.m.f. the current will be continually changing and an alternating back e.m.f. will be self induced.

It has also been seen that when a capacitor is connected in series with a resistor across a d.c. supply the p.d. across the capacitor plates increases gradually. When the capacitor is fully charged the charging current falls to zero. Now, if a capacitor is connected into an alternating current circuit the voltage across the capacitor will be continually changing and hence the charging and discharging current will also continually vary.

The effects of inductance and capacitance in an a.c. circuit occur as long as the circuit is connected to a supply and not only when a switch is closed or opened as in a d.c. circuit. This may appear to make a.c. circuits look more complicated than d.c. circuits but do not let this deter you, for the understanding of such circuits is much easier than might at first be expected. In this chapter we will deal with circuits containing pure resistance, inductance and capacitance, and then with series circuits containing combinations of these quantities.

Resistance in an alternating current circuit

Let us begin with an experiment in which a resistor is connected across a variable frequency a.c. supply.

Experiment 2(i) *To investigate the current through a resistor when the magnitude of the applied voltage is maintained constant and the frequency varied*

In this experiment a 4-pole alternator was driven by a variable speed d.c. motor. The alternator had three windings on the rotor and the output from one winding was connected to supply an external circuit which included a pure resistor and indicating instruments as in Fig. 2.1. A pure resistor is wound non-inductively and the effect of capacitance between the turns is negligible.

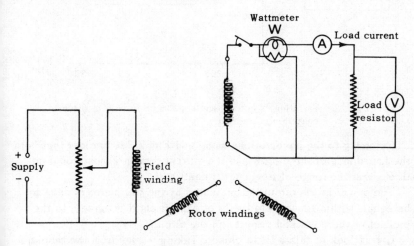

Fig. 2.1. Resistance in an a.c. circuit.

Throughout the test the alternator field current was adjusted to maintain 100 volts across the load resistor which had a nominal value of 175 ohms. The alternator speed was varied in steps from 900 rev/min to 2400 rev/min and the instrument readings observed at each stage. Results were as tabulated.

P.D. across R, V (V)	100	100	100	100	100	100
Current, I (A)	0·58	0·575	0·575	0·57	0·57	0·57
Speed (rev/min)	900	1200	1500	1800	2100	2400
Frequency (Hz)	30	40	50	60	70	80
Power (W)	57	57	57	56	56	56·5
Impedance, $Z = V/I$ (Ω)	172	174	174	175	175	175
Voltamperes, $V \times I$ (VA)	58	57·5	57·5	57	57	57

In an alternating current circuit the ratio V/I is termed the impedance (Z) and measured in ohms. It may be looked upon as the opposition which the circuit presents to the flow of current through it.

The product $V \times I$ in an alternating current circuit has the units of voltamperes. We already know that in a d.c. circuit the product $V \times I$ gives us the power in watts but as we shall see later this is not necessarily true in an a.c. circuit.

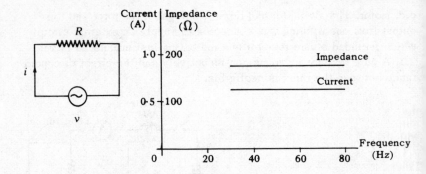

Fig. 2.2. Variation of current and impedance in an a.c. circuit containing resistance.

Referring to the observations made and Fig. 2.2 it is clear that both the impedance of the circuit and the current remain almost constant throughout the range of frequencies dealt with.

The power in the circuit as indicated by the wattmeter is seen to be equal, within reasonable limits of experimental accuracy, to the product of the volts and amperes in the circuit.

Let us look at this a little closer. Taking values from the tabulated results when the frequency of the alternator output is 50 Hz it is seen that there is a p.d. of 100 volts across the resistor and a current of 0·575 A through it. These are r.m.s. values and converting to maximum values we have 142 volts and 0·813 ampere respectively.

Figure 2.3 includes the voltage and current waves. The current at any instant is given by $i = \nu/R$ and in a circuit containing pure resistance the voltage and current waves are in phase.

The phasors **V** and **I** are coincident but for clarity are shown separated. A curve of power is also shown. Ordinates of this power curve are obtained by taking the products of instantaneous values of ν and i. The power is seen to fluctuate and has a mean value of 57·5 watts which agrees with the wattmeter indication.

We may verify this mean power mathematically.

Let the applied voltage, $\nu = V_m \sin \omega t$

then the resulting current $i = I_m \sin \omega t$

The power at any instant $= \nu \times i$

$[2\sin^2 \omega t = 1 - \cos^2 \omega t]$ $= V_m I_m \sin^2 \omega t = \frac{1}{2} V_m I_m (1 - \cos 2\omega t)$

$$= \frac{V_m I_m}{2} - \frac{V_m I_m \cos 2\omega t}{2}$$

= constant component − fluctuating component.

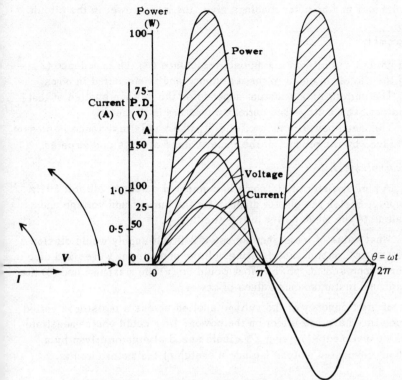

Fig. 2.3. Voltage, current, and power waves for an a.c. circuit of pure resistance.

Now, the mean power will be given by the average value of this expression. The expression may be identified in Fig. 2.3 as a constant component represented by the broken line $A - A$, together with a component of maximum value 57·5 watts which alternates about the broken line. This alternating component has a frequency twice that of the constituent voltage and current waves.

The average value of the fluctuating component is zero and hence the mean power $= V_m I_m /2$

$$= \frac{V_m}{\sqrt{2}} \times \frac{I_m}{\sqrt{2}}$$

For a circuit containing pure resistance the mean power $= V \times I$ where V and I are r.m.s. values.

Since the instruments used in this experiment are normal laboratory instruments calibrated in terms of r.m.s. values the product of the voltmeter and ammeter readings gives the mean power in the circuit.

Summary

In an a.c. circuit containing only resistance (R) the impedance (Z) of the circuit is equal to the resistance and is measured in ohms.

The current at any instant depends on the voltage applied at that instant. The voltage and current waves are in phase.

The mean power in an a.c. circuit, where only resistance is present, is given by the product of the r.m.s. values of volts and amperes.

Examples 2.1.

1. A pure resistor of 100 ohms is connected across a 240 volt 50 Hz supply. Sketch the phasor diagram for the current and voltage and calculate the mean power in the circuit.

2. What value of resistance across a 240 volt supply would dissipate 500 watts? Sketch the resulting voltage and current waves and also a curve representing power. What would be (a) the maximum and (b) the minimum instantaneous values of power?

3. If the frequency of the voltage applied across a resistor is varied does this have any effect on the power? How could you demonstrate this experimentally using a variable speed alternator driven by a shunt connected motor? Include a sketch of the motor circuits.

Inductance in an alternating current circuit

In the previous experiment, we considered an alternating current circuit with only resistance. Very often however inductance may also be present. Let us now consider a highly inductive a.c. circuit and, for a time, neglect the effect of any resistance.

Experiment 2(ii) *To investigate the current and power in an inductive circuit when the magnitude of the applied voltage is maintained constant and the frequency is varied*

The same alternator was used as in the previous test. An inductive coil with a laminated iron core and a winding of negligible resistance, was connected across the alternator output terminals (Fig. 2.4).

The speed of the motor driving the alternator was varied in steps. The magnitude of the variable frequency output was maintained constant at 80 volts by adjusting the alternator field regulator. The current through the coil and the power in the circuit were measured at different frequencies. The results were as tabulated.

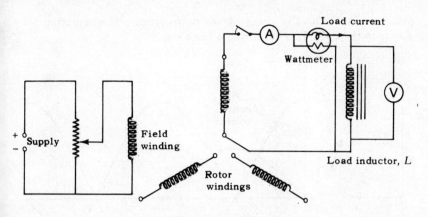

Fig. 2.4. Inductance in an a.c. circuit.

P.D. across L, V(V)	80	80	80	80	80	80	80	80	80
Current, I (A)	1·05	0·92	0·76	0·66	0·57	0·5	0·46	0·41	0·39
Speed (rev/min)	600	750	900	950	1200	1350	1500	1650	1800
Frequency, f (Hz)	20	25	30	35	40	45	50	55	60
Power (W)	0	0	0	0	0	0	0	0	0
Impedance, $Z = V/I$ (Ω)	76	87	106	121	141	160	174	195	205
Voltamperes, $V \times I$ (VA)	84	74	61	53	46	40	37	33	31

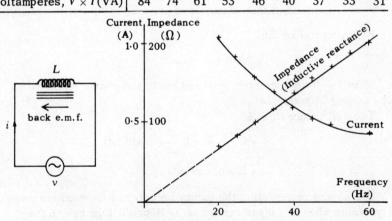

Fig. 2.5. Variation of current and inductive reactance in an a.c. circuit as the supply frequency is varied.

The power in the circuit as indicated by the wattmeter is zero. It is *not* equal to the product of the voltmeter and ammeter readings.

Let us look closer at a circuit containing pure inductance. The current through the coil is alternating and consequently the flux it produces is continually changing and self-inducing an alternating e.m.f. in the coil. This is termed a back e.m.f.

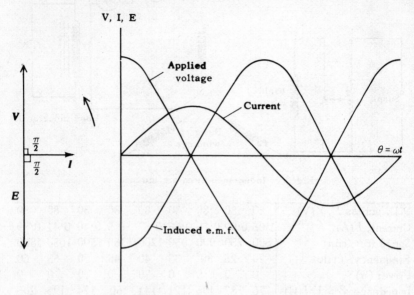

Fig. 2.6. Applied voltage, current and induced e.m.f. waves for a purely inductive circuit.

Referring to Fig. 2.6.

$$\text{Let } i = I_m \sin \omega t$$

When $t = 0$ the current is zero and hence the phasor representing the current is horizontal.

The self induced e.m.f.

$$e = -L\frac{di}{dt} = -L\frac{d}{dt}(I_m \sin \omega t)$$

$$e = -L\omega I_m \cos \omega t$$

The phasor representing the induced e.m.f. has its maximum negative value when $t = 0$ and $\cos \omega t = 1$. Hence it lags behind the current phasor by $\pi/2$ radians.

To produce a current the applied voltage must be equal and opposite to the self induced e.m.f.

$$\therefore v = +L\omega I_m \cos \omega t$$
$$= L\omega I_m \sin(\omega t + \pi/2)$$

The phasor representing this applied voltage wave has its maximum positive value when $t = 0$. Hence this phasor leads the current phasor by $\pi/2$ radians. Stated another way, we may say that in a purely inductive circuit the current lags behind the applied voltage by $\pi/2$ radians.

Now in the last equation for an alternating voltage wave the term $L\omega I_m$ will have the units of volts and be equal to V_m.

We may therefore write,

$$\nu = V_m \sin(\omega t + \pi/2) \text{ where } V_m = L_w I_m$$

since $L\omega$ is equal to the ratio volts/amperes the units which it is expressed will be ohms.

$L\omega = 2\pi fL$ is termed the inductive reactance of the coil and denoted by X_L. It may be thought of as the opposition which the coil presents to the flow of current through it, because of the back e.m.f. induced as the current alternates. Any resistance in the coil would also offer opposition to current flow. With the coil used the resistance was negligible.

Inductive reactance $X_L = 2\pi fL$ is thus seen to be directly proportional to the frequency of the supply, and this is confirmed by the experimental results as plotted in Fig. 2.5.

Power in a purely inductive circuit

The zero indication of the wattmeter in this experiment is rather interesting. We have a voltage applied and current flowing but apparently no power!

Let us consider carefully the experimental results obtained when the applied frequency is 50 Hz. It is seen that the r.m.s. values of the current and the applied voltage are 0·46 ampere and 80 volts. The corresponding maximum values of these waves will therefore be 0·65 ampere and 113 volts respectively.

Figure 2.7 shows the applied voltage and current waves, and also a curve of power plotted by considering the products of the instantaneous related values of current and voltage.

The power curve is seen to fluctuate about the horizontal axis and the mean value is zero as confirmed by the wattmeter reading.

Let us verify this power curve mathematically. From the phasor diagram Fig. 2.7.

$$i = I_m \sin \omega t$$
$$\text{then } \nu = V_m \sin(\omega t + \pi/2)$$
$$= V_m \cos \omega t$$

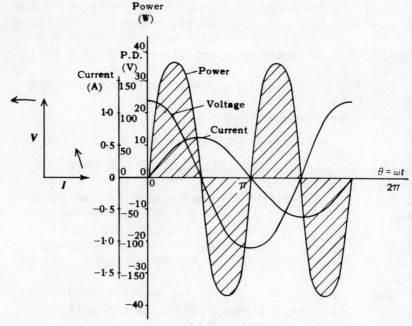

Fig. 2.7. Voltage, current and power waves for a purely inductive circuit.

The power at any instant

$$= v \times i$$

$[\sin 2\alpha t = 2\sin\alpha t \cos\alpha t]$

$$= V_m I_m \sin \omega t \cos \omega t$$

$$= \frac{V_m I_m}{2} \sin 2\omega t$$

This is an alternating power wave which has twice the frequency of the constituent voltage and current curves.

It has a maximum value of $V_m I_m / 2$ equal to 36·7 watts and a mean value of zero.

Summary

In an alternating current circuit containing pure inductance:

(a) the impedance of the circuit is equal to the inductive reactance (X_L) measured in ohms and $X_L = 2\pi f L$.

(b) as the frequency of the supply increases the inductive reactance increases and the current decreases.

(c) the current lags $\pi/2$ radians behind the applied voltage.

(d) the mean power in the circuit is zero.

Examples 2.2.

1. Explain why the current in an a.c. circuit containing only pure inductance lags $\pi/2$ radians behind the applied voltage. Sketch curves representing the current, induced e.m.f. and applied voltage for such a circuit.

2. What is meant by inductive reactance? A purely inductive coil has an inductance of 0·318 H and is connected across a 100 volt 50 Hz supply. Plot curves showing how the applied voltage, current and power vary during one cycle of the applied voltage. Write down in relation to your diagram the equation of the power wave.

3. How does inductive reactance vary with frequency? Sketch a curve showing how the current through a 0·1-H inductor would vary if an alternating voltage of maximum value 50 volts and of variable frequency was applied to the inductor. Consider a frequency range from 10 Hz to 100 Hz.

Resistance and inductive reactance connected in series

Circuits rarely consist of pure resistance or pure inductive reactance and we must know how to deal with cricuits which include both resistance and reactance.

Experiment 2(iii) *To investigate the current, voltage drops and power in a series circuit containing resistance and inductive reactance.*

In this experiment a resistor (R) with negligible reactance and a coil (L) of almost pure inductance were connected in series across a 150 volt 50 Hz supply (Fig. 2.8).

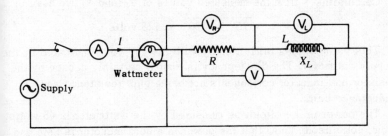

Fig. 2.8. Resistance and inductive reactance in connected series.

The instrument readings were as tabulated.

P (W)	I (A)	V_R (V)	V_L (V)	V (V)
45	0·59	76	126	150

Let us draw the phasor diagram for this circuit. We start by drawing the phasor representing the quantity common to all components in the circuit. In a series circuit this is the current phasor (Fig. 2.9.)

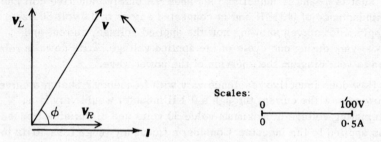

Fig. 2.9. $R - L$ Series circuit phasor diagram.

Now, V_R across the resistor is in phase with the current I. With a pure inductance in a circuit the current lags $\pi/2$ radians behind the applied voltage and we may therefore draw V_L so that I lags $\pi/2$ radians behind V_L.

The total voltage V applied to the circuit must be equal to the phasor sum of V_R and V_L.

Hence

$$V^2 = V_R^2 + V_L^2$$

The supply current will lag behind the supply voltage by an angle ϕ which is between 0 and $\pi/2$ radians, depending on the relative amounts of resistance and inductive reactance in the circuit.

Calculating V from the measured values of V_R and V_L we have

$$V = \sqrt{76^2 + 126^2} = 148 \text{ volts.}$$

This value agrees fairly closely with the 150 volts indicated by the supply voltmeter. The discrepancy is almost certainly caused by the resistor and inductor components not being pure resistance and inductance units.

The power in the circuit as measured by the wattmeter is 45 watts. Now, you already know that the power in a pure inductor is zero and hence the only power in the circuit is that in the resistor. Can we calculate the power in the resistor?

$$\text{Power in the resistor} = V_R \times I$$
$$= V\cos\phi \times I$$
$$\therefore \text{power} = VI\cos\phi$$

In this experiment ϕ as measured from the diagram is 59° and hence $\cos\phi$ is 0·51.

The power, $VI\cos\phi = 150 \times 0·59 \times 0·51 = 45·2$ watts. This agrees very closely with the wattmeter reading.

Figure 2.10 shows the supply voltage, current and power waves and also the mean power in the circuit. The maximum values of the supply voltage and current waves are $150 \times \sqrt{2}$ volts and $0·59 \times \sqrt{2}$ amperes, giving 212 volts and 0·834 amperes respectively.

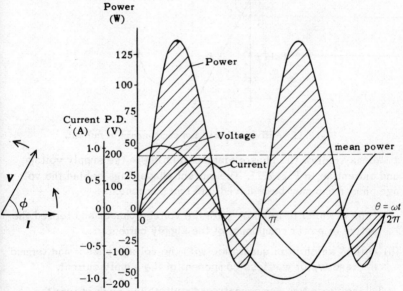

Fig. 2.10. Voltage current and power waves for a series circuit containing resistance and inductive reactance.

True power, apparent power and power factor

Power in an alternating current circuit as given by $V \times I \times \cos\phi$ is termed true power and is measured in watts. This is the quantity indicated by a wattmeter in an a.c. circuit.

The product $V \times I$ is often termed the apparent power and is measured in voltamperes.

Now, in an a.c. circuit the ratio,

$$\frac{\text{true power}}{\text{apparent power}}$$

is termed the power factor of the circuit.

For sine waves,

$$\text{power factor} = \frac{\text{true power}}{\text{apparent power}} = \frac{VI\cos\phi}{VI} = \cos\phi$$

Active and reactive current components

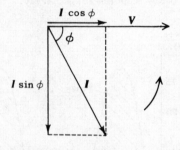

Fig. 2.11. Active and reactive components of current.

We may redraw the phasor diagram relating to the supply voltage and current as in Fig. 2.11. The current phasor lags behind the voltage phasor and may be resolved into two components:

(a) $I\cos\phi$ which is in phase with the voltage phasor and termed the active or power component of the supply current.

(b) $I\sin\phi$ which is in quadrature with the voltage phasor and termed the reactive or wattless component of the supply current.

Taking these two components and multiplying each in turn the supply voltage V we obtain $V \times I\cos\phi$ watts (W) or true power and $V \times I\sin\phi$ reactive voltamperes (VAr).

Impedance triangle for R and L in series

The ratio supply voltage/supply current gives us the impedance of an a.c. circuit and is measured in ohms. In the circuit of the preceding experiment it is obvious that the impedance is a combination of resistance and inductive reactance.

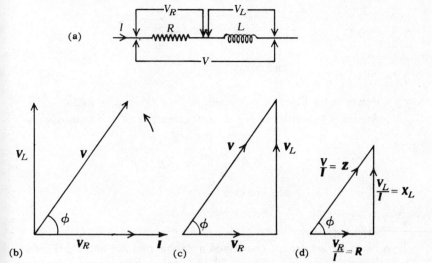

Fig. 2.12. Impedance triangle for a series circuit of resistance and inductive reactance.

Now, referring to the circuit of Fig. 2.12(a) you will observe that the phasor diagram may be drawn as in Fig. 2.12(b). In Fig. 2.12(c) the three voltage phasors are shown forming a closed triangle. Since the same current flows throughout the series circuit we may divide the voltages represented by each side of the triangle by the current and obtain a similar triangle as in Fig. 2.12(d), where the sides of the triangle represent resistance R, inductive reactance X_L, and impedance Z.

This is known as the impedance triangle for the series circuit and it is clear from the diagram that **Z** = **R** + **X**$_L$ or,

$$Z^2 = R^2 + X_L^2$$

This is an important expression which enables us to combine resistance and inductive reactance.

Also,

$$\cos \phi = \frac{R}{Z}$$

Summary

In a series circuit containing resistance and inductive reactance the current lags behind the applied voltage by an angle of between 0 and $\pi/2$ radians.

$$V = V_R + V_L$$
$$\therefore V^2 = V_R^2 + V_L^2$$
$$\text{and } \cos\phi = \frac{V_R}{V}$$

Power in an $R - L$ a.c. circuit $= V \times I \times \cos\phi$ watts.
Apparent power in an $R - L$ a.c. circuit $= V \times I$ voltamperes

$$Z = R + X_L$$
$$\therefore Z^2 = R^2 + X_L^2$$
$$\text{and } \cos\phi = \frac{R}{Z}$$

Examples 2.3.

1. An a.c. series circuit comprises a 25 ohm resistor and 0·1 H inductor connected across a 100 volt 50 Hz supply. Draw a phasor diagram representing the supply voltage and current, and also the voltage drops across the circuit components.

2. How is the impedance triangle derived for an a.c. series circuit with resistance and inductive reactance? Determine the total impedance of a circuit which comprises a resistor of 2·2 kΩ connected in series with an inductor of 2 mH when the frequency of the supply is 100 kHz.

3. A voltage represented by $v = 250 \sin(628t + \pi/3)$ produces a current $i = 50 \sin 628t$ when applied across a coil. What are the frequency, maximum and r.m.s. values of the voltage wave and what does $\pi/3$ signify? Why is power absorbed when an alternating current flows in a resistor even though the current is alternatively positive and negative?

4. A coil of wire dissipates 50 watts when carrying a direct current of 2 amperes. What is the resistance of the coil? When an alternating voltage of 100 volts is applied the current is again 2 amperes. What is the impedance of the coil? Why is the impedance higher than the resistance?

5. Explain the meaning of power factor. A single phase a.c. motor with an output of 4 hp and efficiency of 88% takes an input current on full load of 20 amperes from a 200 volt supply. What is the power factor of the load? (1 hp = 746 W).

6. Explain what is meant by efficiency, power factor and impedance. An electric motor develops a torque of 6 newton metres at a speed of 1250 rev/min and takes a current of 6·0 A from a 220 V 50 Hz supply.

The power input is 1000 W. Calculate the power factor and efficiency of the motor. What would be the impedance of a coil which would pass the same current as the motor when connected to the same supply?

7. Why is it more economical to use a reactor instead of a resistor to dim a bank of lamps? If a bank of lamps takes 4 kW at 250 V determine the impedance of a reactor required to reduce the lamp voltage to 150 V assuming that the resistance of the lamps does not change.

8. A coil having 1·59 mH inductance and 10 ohms resistance is connected in turn to the following sources of e.m.f. (a) 12 volts d.c. (b) 12 volts 400 Hz. What value of current flows through the coil in each case? Explain why the currents differ.

9. A single phase 240 volt a.c. motor is required to drive a pump which raises 1000 litres of water per minute to a height of 25 metres. If the combined efficiency of the motor and pump is 70% calculate the power input to the motor and also the motor current when the operating power factor is 0·7. Sketch a circuit diagram showing how a wattmeter may be connected to measure the input power.

10. Show that with a sinusoidal waveform the average power in an a.c. circuit is given by the product of the r.m.s. values of the current, and voltage and the cosine of the angle between the voltage and current phasors.

An electric motor takes a current, which has a maximum value of 8 amperes, from an alternating supply of maximum value 250 volts. The voltage wave leads the current wave by $\pi/6$ radians.

Plot the voltage and current waves over one complete cycle and also a curve representing power.

11. An alternating current electromagnet coil dissipates 500 W when carrying a current of 10 A. Calculate the voltage across the coil when it is connected in series with a non-reactive resistor of 15 ohms across a 240 V 50 Hz supply, the current being 10 A.

12. When a 5 volt d.c. supply is applied across a certain circuit the current flowing is 0·1 ampere. When the d.c. supply is replaced by a sinusoidal a.c. supply with a frequency of 80 Hz it is found that a voltage of 10 volts is necessary to produce a current of 0·1 ampere in the same circuit. Give reasons for the difference in voltage and calculate the d.c. resistance of the circuit, the power absorbed under the d.c. conditions, and the impedance of the circuit at 80 Hz.

13. A single-phase industrial load is equivalent to a resistor of 12 ohms in series with an inductor of 9 ohms reactance. The supply lines have a combined resistance of 0·5 ohm and an inductive reactance of

1·5 ohm. The output voltage of the alternator supplying the load is 6600 volts. Calculate the power factor at which the alternator is operating, the voltage across the terminals of the load and the impedance drop in the line. Sketch a phasor diagram.

14. A 50 Hz circuit has a resistance of 20 ohms, an inductance of 0·07 H and carries a current of 10 A. Determine (a) the maximum rate of change of current in the circuit, (b) the instantaneous voltage required to produce this maximum rate of change of current and (c) the reading of a voltmeter connected across the circuit.

15. Two air-cored inductance coils when connected in series across a direct current supply take a current of 4 amperes and the voltages across the two coils are 60 volts and 80 volts respectively. When connected across an alternating current supply of frequency 25 Hz the current is 4 A but the voltages are now 120 volts and 200 volts respectively. If the coils still connected in series, are now supplied from a 250 volt 50 Hz supply, calculate the total current, the power factor, and the voltage across each coil. Draw to scale a phasor diagram for the final circuit showing the various voltage and current phasors.

16. An air-cored coil is connected in series with a non-inductive resistor across a 250 volt 50 Hz supply. The voltage drops across the coil and resistor are respectively 160 volts and 130 volts when the current is 10 amperes. Sketch a phasor diagram representing the voltage drops and current in the circuit. Calculate the resistance and the inductance of the coil, also its impedance and power factor.

17. A coil is connected through a switch to the terminals of a 110 volt battery. At the instant when the switch is closed the current grows at the rate of 400 A/s. The final steady current is 2 amperes. Sketch the curve of current growth. Determine the current the coil will take if it is permanently connected to an alternating source of 110 volts at 50 Hz.

Capacitance in an alternating current circuit

Now that we have seen how to deal with resistance, and inductive reactance in series alternating current circuits we come to consider capacitance in these circuits.

You are already familiar with capacitance in a d.c. circuit, and having carried out experiments which involve the charging and discharging of capacitors you will appreciate that if a capacitor is connec- across an alternating voltage supply then it will be charged and discharged repeatedly depending on the frequency of the supply.

Let us consider a circuit assumed to contain pure capacitance.

Experiment 2(iv) *To investigate the current and power in a circuit of pure capacitance, when the magnitude of the applied voltage is maintained constant and the frequency varied*

A capacitor of nominal value $15\,\mu\text{F}$ was connected across the output terminals of an alternator driven by a variable speed d.c. motor (Fig. 2.13).

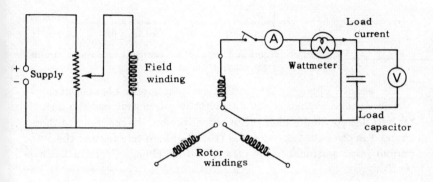

Fig. 2.13. Capacitance in an a.c. circuit.

The speed of the driving motor was varied in steps to produce a variable frequency alternator output voltage. With the alternator field current adjusted to maintain the magnitude of the output voltage constant at 120 volts, the current through the capacitor and the power in the circuit were measured at different frequencies.

The observations were tabulated as follows:

P.D. across C, V (V)	120	120	120	120	120	120	120	120	120	120
Current, I (A)	0.39	0.45	0.5	0.57	0.63	0.7	0.76	0.82	0.86	0.90
Speed (rev/min)	1050	1200	1350	1500	1650	1800	1950	2100	2250	2400
Frequency, f (Hz)	35	40	45	50	55	60	65	70	75	80
Power (W)	0	0	0	0	0	0	0	0	0	0
Impedance, $Z = V/I$ (Ω)	308	267	240	210	192	173	159	146	139	133
Voltamperes, (VA)	47	54	60	68	76	84	91	98	103	108

The results of the experiment show (Fig. 2.14) that the impedance of the capacitor falls as the frequency increases, and consequently the current in the circuit rises. The power in the circuit as indicated by the wattmeter appears to be zero.

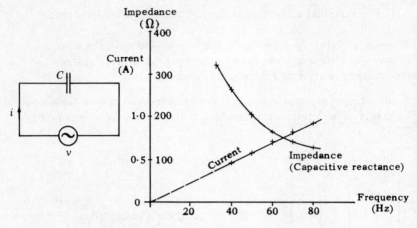

Fig. 2.14. Variation of current and capacitive reactance in an a.c. circuit as the supply frequency is varied.

Let us look closer at this circuit containing pure capacitance.

The voltage applied to the capacitor is alternating and the capacitor will be repeatedly charged in one direction, discharged and then charged in the opposite direction (Fig. 2.15). At any instant the current is proportional to the rate of voltage change (Volume 1. Chapter 7).

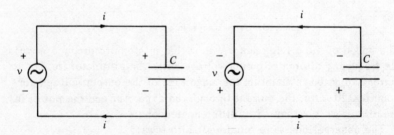

Fig. 2.15. Charge and discharge of a capacitor.

Referring to Fig. 2.16 let,

$$v = V_m \sin \omega t$$

$$\therefore i = C \frac{d}{dt}(V_m \sin \omega t)$$

$$\therefore i = \omega C V_m \cos \omega t$$

When $t = 0$ the instantaneous current will be $\omega C V_m$ and hence the phasor representing the current leads the applied voltage by $\pi/2$ radians.

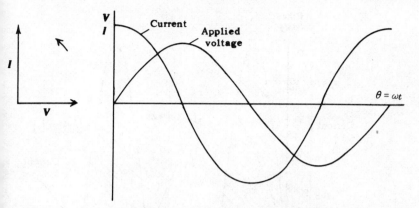

Fig. 2.16. Alternating current circuit with pure capacitance.

In a purely capacitive circuit the current leads the voltage by $\pi/2$ radians.

The maximum value of the current, $I_m = \omega C V_m$ hence

$$\frac{V_m}{I_m} = \frac{1}{\omega C} \text{ ohms}$$

$$X_c = \frac{1}{\omega C} = \frac{1}{2\pi f C}$$

is termed the capacitive reactance of the circuit. It may be looked upon as the opposition which the capacitor presents to the flow of current in the circuit.

Capacitive reactance is inversely proportional to the frequency of the supply and this is confirmed by the graph (Fig. 2.14).

Power in a purely capacitive circuit

The power in the circuit as indicated by the wattmeter is zero. Let us look at this in more detail and consider the experimental results when the frequency of the applied voltage is 50 Hz. The r.m.s. values of the applied voltage and current are 120 volts and 0·57 ampere. Corresponding maximum values which enable us to draw the waves are 170 volts and 0·8 ampere respectively.

Figure 2.17 shows the applied voltage and resulting current waves. The power curve is obtained by considering the instantaneous products of corresponding values of current and voltage.

The power curve is seen to fluctuate about the horizontal axis and the mean power as confirmed by the wattmeter indication is zero.

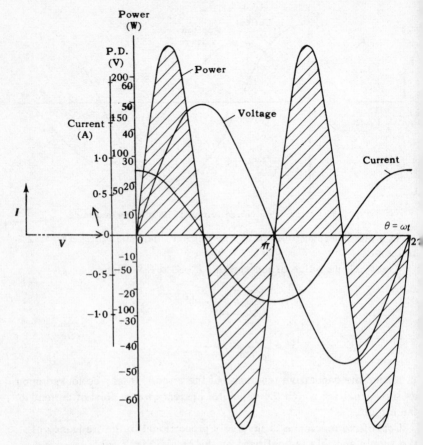

Fig. 2.17. Voltage, current and power curves for a purely capacitive circuit.

We may verify this mathematically.

$$\text{If } \nu = V_m \sin \omega t$$
$$\text{then } i = I_m \cos \omega t$$

and the power at any instant,

$$= \nu \times i$$
$$[\sin 2\omega t = 2\sin \omega t \cos \omega t] \quad = V_m I_m \sin \omega t \cos \omega t$$
$$= \frac{V_m I_m}{2} \sin 2\omega t$$

This is an alternating power wave which has twice the frequency of the constituent voltage and current waves. In this experiment as it

56

has a maximum value of

$$\frac{V_m I_m}{2} = 68 \cdot 0 \text{ watts}$$

and a mean value of zero.

Summary

In an alternating current circuit containing pure capacitance:

(a) the impedance of the circuit is equal to the capacitive reactance (X_c) measured in ohms.

$$X_c = \frac{1}{2\pi f C}$$

(b) as the frequency of the supply is increased the capactive reactance decreases and the current in the circuit increases.
(c) the current leads the applied voltage by $\pi/2$ radians.
(d) the mean power in the circuit is zero.

Examples 2.4.

1. Explain why the current in a circuit containing only pure capacitance leads the applied voltage by $\pi/2$ radians.

If a voltage of maximum value 100 volts at 50 Hz is applied to a 100μF capacitor sketch curves representing the voltage and current waves.

2. How does capacitive reactance vary with frequency? A 10μF capacitor is connected across an alternating voltage of maximum value 50 volts. If the frequency is varied over the range 10 Hz to 1000 Hz plot a curve of current against frequency.

3. The current in a purely capacitive circuit has a maximum value of 5 amperes and the maximum value of the applied sinusoidal voltage wave is 1000 volts. Sketch the current and voltage waves and also a curve representing power in the circuit. Express the power curve by an equation and determine the average power in the circuit.

4. Explain how capacitance can be measured using an a.c. ammeter and voltmeter.

Determine the capacitance of each of two similar capacitors which take a current of 1 A when connected in series with each other across a 200 V 50 Hz supply.

Resistance and capacitive reactance connected in series

We now come to the study of circuits containing resistance and capacitive reactance connected in series.

Experiment 2(v) *To investigate the current, voltage drops and power in a series circuit containing resistance and capacitive reactance*

In this experiment a resistor (R) and a capacitor (C) were connected in series across a 100 volt 50 Hz supply (Fig. 2.18).

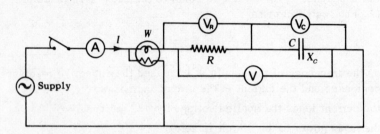

Fig. 2.18. Resistance and capacitive reactance in series.

The instrument readings were observed and tabulated.

P (W)	I (A)	V_R (V)	V_C (V)	V (V)
33·5	0·98	35	94	100

Let us draw the phasor diagram for the circuit by commencing with the current phasor since the current is common to all components in a series circuit.

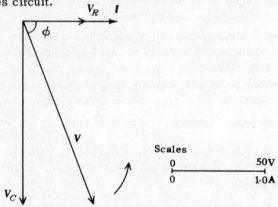

Fig. 2.19. Series circuit phasor diagram.

58

Now, the voltage drop V_R across the resistor will be in phase with the current. In a circuit containing pure capacitance the current leads the voltage across the capacitor by $\pi/2$ radians. We may therefore draw V_c as shown.

The total voltage V applied to the circuit must be equal to the phasor sum of V_R and V_c.

Hence
$$V^2 = V_R^2 + V_c^2$$

The supply current in this circuit leads the supply voltage by an angle ϕ of between 0 and $\pi/2$ radians depending on the relative amounts of resistance and capacitive reactance in the circuit.

Calculating V from the measured values of V_R and V_c we have
$$V = \sqrt{35^2 + 94^2} = 100 \text{ volts.}$$

This value agrees with the 100 volts indicated by the supply voltmeter.

The power in the circuit as measured by the wattmeter is 33·5 watts. Now the power in a pure capacitor is zero and hence the only power in the circuit is the power in the resistor.

Power in the resistor = $V_R \times I = V \cos\phi \times I = VI \cos\phi$.

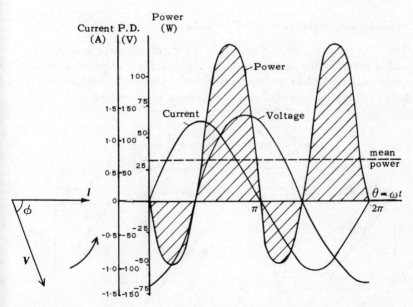

Fig. 2.20. Voltage, current and power waves for a series circuit containing resistance and capacitive reactance.

From the phasor diagram (Fig. 2.19)
$$\cos\phi = \frac{V_R}{V} = 0\cdot 35$$
$$\therefore \text{power} = 100 \times 0\cdot 98 \times 0\cdot 35 = 34\cdot 2 \text{ watts.}$$

This value agrees closely with the wattmeter reading.

Figure 2.20 shows the supply voltage, current and power waves, and also the mean power in the circuit. The maximum value of the voltage and current waves are $100 \times \sqrt{2}$ and $0\cdot 98 \times \sqrt{2}$ amperes giving 141 volts and 1·39 amperes respectively.

True power, apparent power, and power factor

Just as in the circuit containing resistance and inductive reactance so the power in this alternating current circuit is given by the product of $V \times I \times \cos\phi$. This is termed the true power and measured in watts. It is the quantity measured by a wattmeter.

The product $V \times I$ is often termed the apparent power and is measured in voltamperes. If we are dealing with pure sine waves then,

$$\text{power factor} = \frac{\text{true power}}{\text{apparent power}} = \frac{VI\cos\phi}{VI} = \cos\phi$$

Impedance triangle for R and C in series

The impedance a circuit presents to the supply is measured in ohms and is given by the ratio, supply voltage/supply current. In a circuit containing resistance and capacitance the impedance will obviously be a combination of resistance and capacitive reactance.

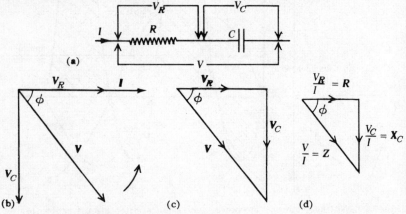

Fig. 2.21. Impedance triangle for a series circuit of resistance and capacitive reactance.

Now, referring to Fig. 2.21(a) you will observe that the phasor diagram may be drawn as in Fig. 2.21(b). The three voltage phasors are shown in Fig. 2.21(c) to form a closed triangle. The same current flows through all components in a series circuit and hence if we divide the sides of this triangle by the current we obtain an impedance triangle as in Fig. 2.21(d) where the three sides of the triangle represent resistance R, capacitive-reactance X_c and impedance Z. It is clear that the total impedance is given by

$$\mathbf{Z} = \mathbf{R} + \mathbf{X}_c$$
$$Z^2 = R^2 + X_c^2$$
$$\text{and } \cos\phi = \frac{R}{Z}$$

Summary

In a series a.c. circuit containing resistance and capacitive reactance the current leads the applied voltage by an angle between 0 and $\pi/2$ radians.

$$\mathbf{V} = \mathbf{V}_R + \mathbf{V}_c$$
$$\therefore V^2 = V_R^2 + V_c^2$$
$$\cos\phi = \frac{V_R}{V}$$

Power in an $R - C$ a.c. circuit $= V \times I \times \cos\phi$ watts.
Apparent power in an $R - C$ a.c. circuit $= V \times I$ voltamperes.

$$\mathbf{Z} = \mathbf{R} + \mathbf{X}_c$$
$$Z^2 = R^2 + X_c^2$$
$$\text{and } \cos\phi = \frac{R}{Z}$$

Examples 2.5.

1. A $100\mu\text{F}$ capacitor and a 30 ohm resistor are connected in series across a 200 volt 50 Hz supply. Give a phasor diagram representing the supply voltage and current and also the voltage drops across the components in the circuit.

2. What is meant by the power factor in a circuit which contains resistance and capacitive reactance connected in series?
   Calculate the value of resistance to be connected in series with a $150\mu\text{F}$ capacitor if the power factor of the circuit is to be 0·5 when the supply frequency is 60 Hz.

3. Explain the impedance triangle for a series circuit of resistance and capacitive reactance. Determine the total impedance of a circuit containing a 100 ohm resistor, a 200 ohm resistor, a $10\mu F$ capacitor and a $20\mu F$ capacitor all connected in series. What is the power factor of the circuit and the voltage drop across the $20\mu F$ capacitor when the circuit is connected across a supply of maximum value 100 volts and the frequency is 50 Hz?

4. Compare the heating effects of a direct current of 2 amperes and a sinusoidal current of peak value 2 amperes flowing in similar resistors. What would be the effect on the heat produced in each circuit if a capacitor was connected in series with the resistor?

Resistance, inductive reactance and capacitive reactance connected in series

Resistance, inductive reactance and capacitive reactance may all be present in a series circuit. Here is an experiment which deals with such a circuit.

Experiment 2(vi) *To investigate the current, voltage drops and power in a series circuit comprising resistance, inductive reactance and capacitive reactance*

A series network of the components was connected across the output terminals of an alternator (Fig. 2.22).

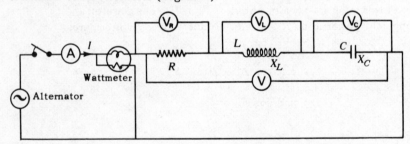

Fig. 2.22. Series R, L, C circuit.

The alternator field current was adjusted until the output was 80 volts at 50 Hz and all the instrument readings were observed.

V (V)	P (W)	I (A)	V_R (V)	V_L (V)	V_c (V)
80	15·4	0·43	35	135	62

Let us draw the phasor diagram for this circuit. We commence with the current phasor which is common to all components in the circuit.

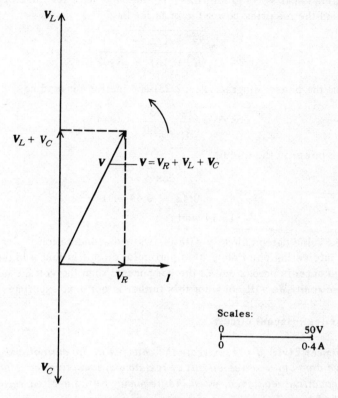

Fig. 2.23. Phasor diagram for R, L and C in series.

The voltage drop across the resistor is in phase with the current while the phasors representing the voltage drops across the inductor and capacitor lead and lag the current phasor by $\pi/2$ radians respectively.

The supply voltage

$$V = V_L + V_c + V_R$$

It is seen from the phasor diagram (Fig. 2.23) that the resulting supply voltage $V = 81\,V$ agrees very closely with the voltmeter reading of 80 volts.

In general when we speak of a power factor being leading or lagging we are referring to the position of the current phasor with respect to the voltage phasor.

If in a series a.c. circuit the magnitude of V_L is greater than V_c the circuit is predominantly inductive and the resulting power factor is lagging but if V_c is greater than V_L the capacitive reactance prevails and the resulting power factor is leading.

$$V = \sqrt{V_R^2 + (V_L - V_c)^2}$$
$$Z = \sqrt{R^2 + (X_L - X_c)^2}$$

From the phasor diagram (Fig. 2.23) and instrument readings,

$$\cos \phi = \frac{V_R}{V} = \frac{35}{80} = 0\cdot 44$$

and the power in the circuit

$$= V \times I \times \cos \phi$$
$$= 80 \times 0\cdot 43 \times 0\cdot 44 \text{ watts}$$
$$= 15\cdot 14 \text{ watts}$$

This value agrees closely with the wattmeter indication.

An interesting point about this particular circuit is that a higher voltage appears across one of the components than the voltage applied to the circuit. We will consider this further in our next experiment.

The series resonant circuit

Experiment 2(vii) *To investigate the changes in the current and voltage drops in a series circuit of resistance, inductive reactance and capacitive reactance, when the frequency of the applied voltage is varied*

In this experiment components and instruments were connected across the output terminals of a variable frequency alternator as in Fig. 2.24.

The magnitude of the supply voltage was maintained at 50 volts throughout the test and the frequency varied over the range 25 Hz to 60 Hz.

The observations were as tabulated.

From the experimental results, curves of current, resistance, capacitive reactance, inductive ractance and impedance were plotted (Fig. 2.25).

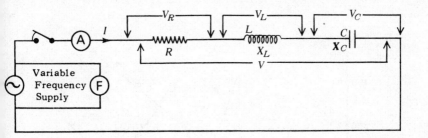

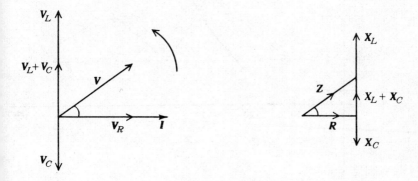

Fig. 2.24. Series R, L, C circuit with phasor and impedance diagrams.

Supply, V	(V)	50	50	50	50	50	50	50	50
Frequency, f	(Hz)	25	30	35	40	45	50	55	60
Current, I	(A)	0·13	0·20	0·29	0·36	0·3	0·25	0·22	0·16
V_L	(V)	28	51	84	130	120	110	103	83
V_R	(V)	13·5	20·1	29	36	30	26	21	15·5
V_c	(V)	76	90	110	115	85	62	53	35
$X_L = V_c/I$	(Ω)	215	255	290	360	400	440	470	520
$R = V_R/I$	(Ω)	104	101	100	100	100	104	96	97
$X_c = V_c/I$	(Ω)	585	450	380	320	285	248	240	218
$Z = V/I$	(Ω)	385	250	163	139	163	200	230	312

It will be observed that as the frequency of the supply is increased the impedance (Z) of the circuit falls to a low value and then increases. At a frequency of slightly below 40 Hz the inductive and capacitive reactances cancel and the impedance of the circuit is limited solely by the circuit resistance (R). The current (I) reaches a maximum value and the voltages across the inductive and capacitive components are considerably higher at this frequency than the voltage applied to the circuit. The circuit power factor is unity.

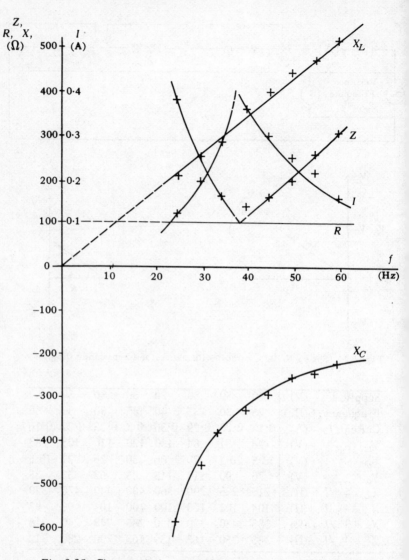

Fig. 2.25. Characteristic curves for a series resonant circuit.

This is an example of resonance in a series a,c. circuit. At resonance very high voltages may be developed across the L and C circuit components.

Resonance occurs when

$$X_L = X_c$$

that is

$$2\pi fL = \frac{1}{2\pi fC}$$

$$\text{or, } f = \frac{1}{2\pi\sqrt{LC}} \text{ Hz}$$

Series circuit magnification

At resonance a series circuit offers low impedance to the passage of current through it. The circuit may be referred to as an acceptor circuit.

The Q factor of the circuit is the voltage magnification, or ratio voltage across L or C to the supply voltage at the resonant frequency.

$$\text{Q factor} = \frac{2\pi fL \times I}{R \times I} = \frac{2\pi fL}{R}$$

$$\text{At resonance } f = \frac{1}{2\pi\sqrt{LC}} \text{ Hz}$$

$$\therefore Q = \frac{1}{R}\sqrt{\frac{L}{C}}$$

Series resonance occurs as a result of an interchange of stored energy between the capacitive and inductive components of the circuit. The rate of energy output from one component is equal in the ideal case to the rate of energy input to the other component, and the circuit does not present any overall reactance. The current from the supply is only limited by the resistance of the circuit.

Summary

Resonance occurs in a series circuit which includes inductive and capacitive reactance, when

$$f = \frac{1}{2\pi\sqrt{LC}} \text{ Hz.}$$

At this frequency the impedance which the circuit presents to the supply is solely the resistance of the circuit and consequently the supply current may be very high.

The supply voltage appears across any resistance in the circuit and voltages much higher than this may appear across the inductive and capacitive components.

The circuit magnification at resonance is given by,

$$Q = \frac{\text{voltage across } L \text{ or } C}{\text{supply voltage}} = \frac{2\pi f L}{R} = \frac{1}{R}\sqrt{\frac{L}{C}}$$

Examples 2.6.

1. A 100 ohm resistor, 0·3 H inductor and 100 μF capacitor are connected in series across a 240 volt 50 Hz supply. Determine the supply current. Give a phasor diagram representing the current and voltage drops in the circuit.

2. A series circuit consisting of a coil of 0·3 H inductance and 15 ohms resistance, and a capacitor of 100 μF is connected across a 240 volt 50 Hz supply. Determine the supply current and the voltages across the coil and capacitor. Draw to scale a phasor diagram representing the supply current and the voltages across all components.

3. A coil having a resistance of 20 ohms with an inductance of 0·05 H is connected in series with a capacitor of 200 μF across a 100 V 50 Hz supply. Determine the voltage across the coil and the power factor of the complete circuit.

4. Explain the meaning of resonance in a circuit containing inductance capacitance, and resistance connected in series. A sinusoidal alternating voltage of 1 volt r.m.s. having a frequency of 1000 kHz is applied across a circuit consisting of a 12·5 mH inductor, a 125 ohm resistor and a capacitor connected in series. Determine the value of C which will produce resonance. Determine also the power dissipated in the circuit and the voltage across the inductor at resonance.

5. A 250 V circuit consisting of a resistor, an inductor and a capacitor connected in series resonates at 50 Hz. The current is then 1 A and the p.d. across the capacitor is 500 V. Calculate the value of the resistor, the inductor, and capacitor.

6. When alternating currents and voltages in a.c. circuits are represented by means of phasor diagrams, what restriction on the wave form is implied? An a.c. supply of 10 volts is connected across a circuit consisting of a 100 ohm resistor and a 0·16 μF capacitor connected in series. Determine the frequency at which the phase angle between the supply voltage and current is $\pi/3$ radian.

An inductor of 0·1 H is now connected in series with the circuit, determine the frequency at which the supply current and voltage will be in phase. What is the magnitude of this current?

7. Derive an expression for the resonant frequency of an alternating current circuit consisting of resistance, inductance and capacitance

connected in series. A circuit consists of a 14 ohm resistor, 0·6 H
inductor and a 150 μF capacitor connected in series across a 280
volt variable frequency supply. At what frequency will the impedance
presented to the supply be a minimum and what will be the magnitude
of the voltage across each component at this frequency? Give a
complete phasor diagram drawn to a suitable scale.

8. A series circuit consists of a 10 ohm resistor, 0·16 H inductor and
a variable capacitor connected across a 110 volt 50 Hz supply. Calculate the value of capacitance to give maximum current in the circuit,
and the voltages across the components under these conditions. What
would be the effect on the power factor of increasing the capacitance?

9. A series circuit consists of an inductor of 1·0 H, a resistor of 30
ohms and a capacitor of 20 μF in series across a 100 volt variable-
frequency supply. Determine the reading of a voltmeter connected
across the combined resistance and inductance components when the
frequency is adjusted to the resonance value.

10. What is meant by the Q-factor of an R, L and C series circuit?
Derive an expression for Q in terms of the circuit component values.
   Determine the Q-factor for a series circuit with resistance of 8 ohms,
inductance of 0·5 H and 4 μF capacitance. At what frequency is the
Q-factor a maximum?

Alternating current circuit circle diagrams

We now come to study the locus or path traced by the current phasor,
and the power expended in $R - L$ and $R - C$ series circuits when the
resistance, inductance or capacitance elements are varied.

Experiment 2(viii) *To investigate the current and power in a series
a.c. circuit of constant resistance and variable inductance*

A circuit was connected as in Fig. 2.26.

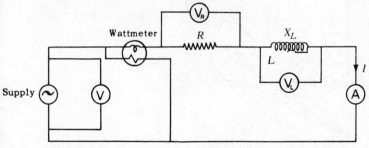

Fig. 2.26. $R - L$ series circuit with variable inductance.

The supply voltage V was maintained at 80 volts throughout the test and by reducing the inductive reactance the current I was increased in steps of one ampere from 3·0 A to 7·0 A. Instrument readings were observed and tabulated at each stage.

V	(V)	80	80	80	80	80
I	(A)	3·0	4·0	5·0	6·0	7·0
V_R	(V)	33	44	56	67	78
V_L	(V)	73	67	57	44	17
P	(W)	102	180	286	407	558
$R = V_R/I$	(Ω)	11	11	11·2	11·1	11·1
$Z = V/I$	(Ω)	26·7	20	16	13·3	11·4
$\cos\phi = R/Z$		0·413	0·55	0·7	0·835	0·975
ϕ	(°)	65·6	56·7	45·5	33·5	13·5

The power factor and phase angle were calculated for each set of readings and the current phasor plotted as shown in Fig. 2.27.

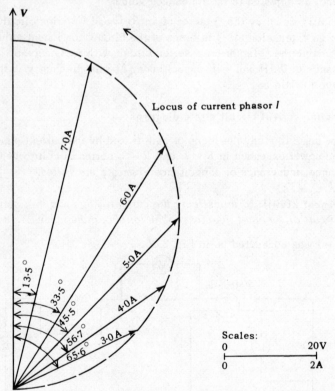

Fig. 2.27. Locus of current phasor.

It is seen that as X_L is varied the locus of the current phasor is a semicircle. How can we account for this?

The supply voltage V produces component voltage drops across R and X_L. The voltage across R will be in phase with the current and the voltage across X_L will be 90° ahead of that across R. For instance when X_L is adjusted so that I is 4 A V_R is 44 V, and V_L is 67 V with relative positions as shown in Fig. 2.28.

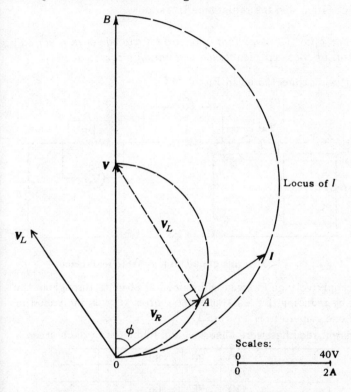

Fig. 2.28. Current and voltage phasor relationships.

V is constant and as V_L and V_R are at right angles to each other other then whatever the value of X_L the point A must lie on a semicircle of diameter V. The current I is given by the ratio V_R/R and hence the phasor V_R represents to an appropriate scale the circuit current. Since the locus of V_R is a semicircle the locus of the current phasor is also a semicircle.

If the reactance was reduced to zero the current would be in phase with the supply voltage, of magnitude V/R and represented by OB.

The power in the circuit is seen to increase as the reactance is reduced.

$$\text{Power} = V \times I \times \cos\phi$$

V is constant at 80 volts and the projection of I (that is $I\cos\phi$) on the axis of the supply voltage increases as the reactance is reduced.

Let us now see what happens when the inductance in the circuit remains constant and the resistance is varied.

Experiment 2(ix) *To investigate the current and power in a series a.c. circuit of constant inductance and variable resistance*

A circuit was connected as in Fig. 2.29.

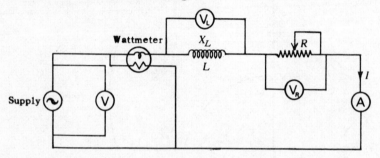

Fig. 2.29. $R - L$ series circuit with variable resistance.

The supply voltage (V) was maintained at 80 volts throughout the test and by reducing the resistance the current (I) was increased in steps of one ampere from 2 A to 5 A.

Instrument readings were observed and tabulated at each stage.

V	(V)	80	80	80	80
I	(A)	2	3	4	5
V_R	(V)	75	66	53	24
V_L	(V)	29	45	60	76
P	(W)	150	198	212	150
$R = V_R/I$	(Ω)	37·5	22	13·25	6
$Z = V/I$	(Ω)	40	26·7	20	16
$\cos\phi = R/Z$		0·94	0·825	0·663	0·375
ϕ	(°)	21	34·5	48·5	68

The power factor was calculated for each set of readings and the current phasor plotted as shown in Fig. 2.30.

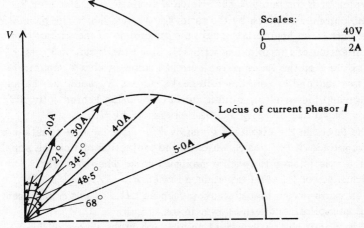

Fig. 2.30. Locus of current phasor.

It is seen that as R is varied the locus of the current phasor is a semicircle. How can we account for this?

The supply voltage V produces component voltage drops across R and X_L. The voltage across R will be in phase with the current and the voltage across X_L will be 90° ahead of that across R. Figure 2.31 shows the conditions when R is adjusted so that the current is 5 A and V_R and V_L have resulting values of 24 V and 76 V respectively.

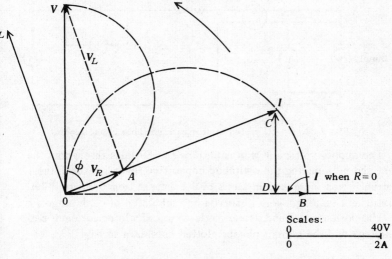

Fig. 2.31. Current and voltage phasor relationships.

73

V is constant and V_L and V_R are at right angles. Thus whatever the value of R the point A must lie on a semicircle of diameter **V**.

The current I is given by the ratio V_L/X_L and hence the phasor V_L represents to an appropriate scale the magnitude of the circuit current. The direction of I is in phase with V_R. Since the locus of V_L is a semicircle then the locus of the current phasor is also a semicircle. If R was reduced to zero the voltage V_L across X_L would be the supply voltage and the resulting circuit current would lag behind V by $90°$, have a magnitude of V/X_L and be represented by OB.

The power in the circuit is given by $V \times I \cos \phi$. V is constant and the magnitude of $I \times \cos \phi$, which is the projection of I on the supply voltage axis is seen to reach a maximum value when $\phi = 45°$. Maximum power is therefore produced when $R = X_L$.

It is worth noting that distances such as $CD = I \cos \phi$ represent to an appropriate scale the power in the circuit.

This circuit has an important application in the theory of the induction motor.

We will now consider $R - C$ circuits.

Experiment 2(x) *To investigate the current and power in an a.c. series circuit of constant resistance and variable capacitance*

A circuit was connected as in Fig. 2.32.

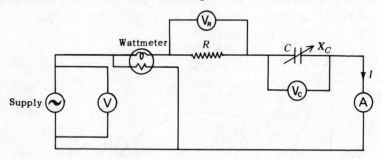

Fig. 2.32. $R - C$ series circuit with variable capacitance.

The supply voltage V was maintained at 80 volts throughout the test and by reducing the switched capacitive reactance bank the current I was increased in steps of $0\cdot1$ ampere from $0\cdot2$ A to $0\cdot7$ A. Instrument readings were observed and tabulated at each stage.

The power factor and phase angle were calculated for each set of readings and the current phasor plotted as shown in Fig. 2.33.

V	(V)	80	80	80	80	80	80
I	(A)	0·2	0·3	0·4	0·5	0·6	0·7
V_R	(V)	20	30	41	51	61	71
V	(V)	77	74	69	62	51	37
P	(W)	5	9	16	23	37	49
$R = V_R/I$	(Ω)	100	100	102	102	102	101
$Z = V/I$	(Ω)	400	267	200	160	133	114
$\cos \phi = R/Z$		0·25	0·375	0·51	0·64	0·77	0·89
ϕ	(°)	76	68	59	50	34	27

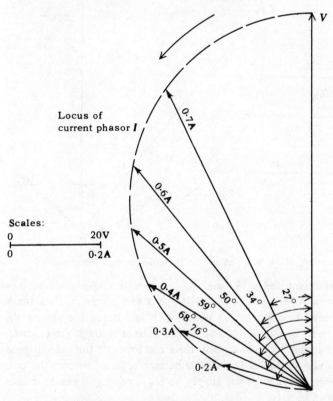

Fig. 2.33. Locus of current phasor.

It is seen that as X_c is varied the locus of the current phasor is a semicircle. How can we account for this?

The supply voltage V produces component voltage drops across R and X_c. The voltage across R will be in phase with the current and the voltage across X_c will be 90° behind that across R. For example, when X_c is adjusted so that I is 0·6 ampere V_R is 61 V and V_c is 51 V as shown in Fig. 2.34.

75

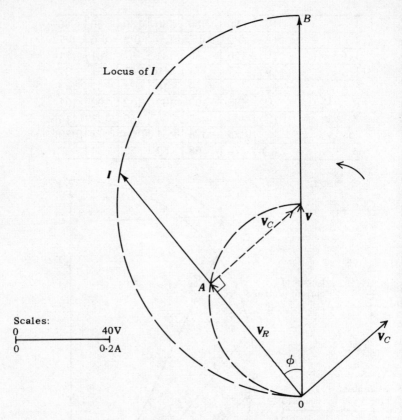

Fig. 2.34. Current and voltage phasor relationships.

V is constant and as V_c and V_R are at right angles then whatever the value of X_c the point A must lie on a semicircle of diameter **V**. The current I is given by the ratio V_R/R and hence the phasor V_R represents to an appropriate scale the current through the circuit. Since the locus of V_R is a semicircle the locus of the current phasor is also a semicircle. If the reactance was reduced to zero the current would be in phase with the supply voltage, of magnitude V/R and represented by OB.

The power in the circuit is observed to increase as the current increases.

$$\text{Power} = V \times I \times \cos \phi$$

V is constant at 80 volts and the projection of I (that is $I \cos \phi$) on the supply voltage axis increases as the reactance is reduced.

Experiment 2(xi) *To investigate the current and power in a series a.c. circuit of constant capacitance and variable resistance*

A circuit was connected as in Fig. 2.35.

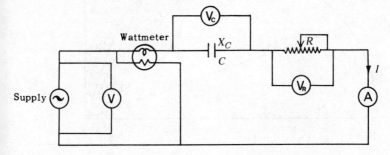

Fig. 2.35. $R - C$ series circuit with variable resistance.

The supply voltage V was maintained at 80 volts throughout the test and, by reducing the resistance, the current I was increased in steps of 0·1 ampere from 0·2 A to 0·6 A.

Instrument readings were observed and tabulated at each stage.

V	(V)	80	80	80	80	80
I	(A)	0·2	0·3	0·4	0·5	0·6
V_R	(V)	76	71	62	49	25
V_c	(V)	25	38	50	63	76
P	(W)	15·1	22	25	24	15
$R = V_R/I$	(Ω)	380	237	155	98	42
$Z = V/I$	(Ω)	400	266	200	160	133
$\cos \phi = R/Z$		0·94	0·9	0·78	0·61	0·31
ϕ	(°)	20	26	39	52	72

The power factor and phase angle were calculated for each set of readings and the current phasor I plotted as in Fig. 2.36.

It is seen that as R is varied the locus of the current phasor is a semicircle. How can we account for this?

The supply voltage V produces component voltage drops across R and X_c. The voltage across R will be in phase with the current and the voltage across X_c will be 90° behind that across R. Figure 2.37 shows the conditions when R is adjusted so that the current is 0·5 A and V_R and V_c have resulting values of 49 V and 63 V respectively.

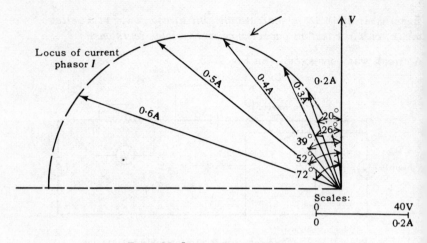

Fig. 2.36. Locus of current phasor.

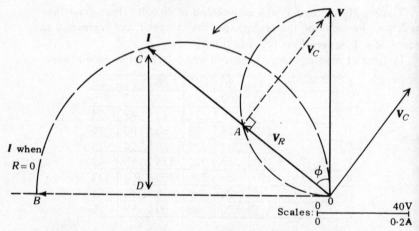

Fig. 2.37. Current and voltage phasor relationships.

V is constant and **V**$_c$ and **V**$_R$ are at right angles. Thus whatever the value of R the point A must lie on a semicircle of diameter **V**.

The current I is given by the ratio V_c/X_c and hence the phasor **V**$_c$ represents to an appropriate scale the magnitude of the circuit current. The direction of I is in phase with **V**$_R$. Since the locus of **V**$_c$ is a semicircle then the locus of the current phasor is also a semicircle. If R was reduced to zero the voltage V_c across X_c would be the supply voltage and the resulting current would lead V by $90°$, have a magnitude of V/X_c and be represented by OB.

The power in the circuit is given by $V \times I \cos \phi$. V is constant and the magnitude of $I \times \cos \phi$, which is the projection of I on the supply voltage axis is seen to reach a maximum when $\phi = 45°$. Maximum power is therefore produced when $R = X_c$.

Distances such as $CD = I \cos \phi$ represent to an appropriate scale the power in the circuit.

Summary

When one component is varied in an $R - L$ or $R - C$ series circuit, the current and component voltage phasors trace semicircles.

As the resistance component is varied the power in the circuit reaches a maximum value when $R = X_L$ or X_c as appropriate.

Examples 2.7.

1. A variable reactor is connected in series with a lamp of 240 ohms resistance, which may be assumed constant. Calculate the reactance range of the reactor required if the lamp current is to be varied over the range 0·4 A to 1·0 A when the supply is 240 V 50 Hz. Sketch a diagram showing the locus of the current phasor.

2. A non-reactive resistor R is variable between 0 and 10 ohms and is connected in series with a coil of resistance 3 ohms and reactance 4 ohms, across a 240 V 50 Hz supply. Sketch a diagram showing the locus of the current phasor when R is (a) zero, (b) 5 ohms and (c) 10 ohms.

3. Show that the locus of the current phasor for a series circuit of fixed capacitance and variable resistance connected across an a.c. supply is a semicircle. If the capacitor has a value of 15·9 μF sketch the current locus as R varies between zero and infinity, when the supply is 250 volts at 50 Hz. What values of current and power factor correspond to maximum power in the circuit and what will be the value of the circuit resistance?

4. A variable capacitor is connected in series with a fixed resistor across a 240 V 50 Hz supply. If the resistor has a value of 30 ohms determine the locus of the end of the current phasor as the capacitor is varied between 225 μF and 75 μF. Calculate the maximum power in the circuit.

3. Resistance, Inductance and Capacitance in Parallel Alternating Current Circuits; The Symbolic Notation; Admittance, Conductance and Susceptance

Introduction

You have studied phasor diagrams for alternating current series circuits and we now come to the study of phasor diagrams for parallel circuits.

In dealing with resistance and reactance in series a.c. circuits it was seen that with the same current flowing through all components impedance triangles could be used in calculations. In parallel circuits however the same current does not flow through all components and impedance triangles cannot be used in the solution of circuit problems involving components connected in parallel.

Initially, we approach parallel circuits by sketching the voltage and current phasor diagrams. Later we will introduce and use a symbolic notational method together with admittance triangles.

Parallel a.c. circuit containing resistance and capacitive reactance

Let us commence our studies by considering an experiment.

Experiment 3(i) *To investigate the currents and power in a parallel network comprising resistance and capacitance connected across an alternating voltage supply*

A parallel circuit was connected across the output of an alternator (Fig. 3.1) which was driven at a constant speed to generate at a frequency of 50 Hz.

The alternator field current was adjusted until the output was 200 volts at 50 Hz. The currents in each branch of the circuit and the total power were observed.

V (V)	I (A)	I_C (A)	I_R (A)	P (W)
200	1·7	1·3	1·1	222

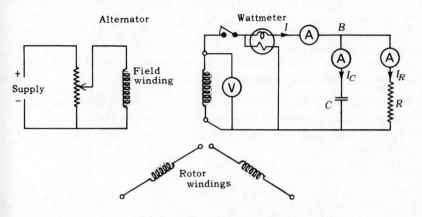

Fig. 3.1. Parallel $R-C$ circuit.

It is seen that the current I flowing towards point B is by no means equal to the arithmetic sum of the currents flowing away from B. It appears at first sight that Kirchhoff's first law does not hold for a.c. circuits.

However, let us sketch the phasor diagram for this parallel circuit remembering to start by drawing the phasor which is common to all components of the circuit. In a parallel circuit the voltage is the common quantity.

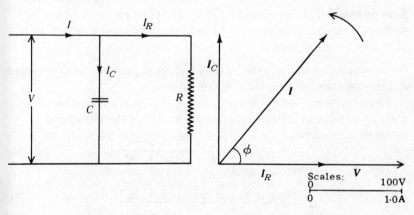

Fig. 3.2. Parallel $R-C$ circuit and phasor diagrams.

The current I_R through the resistor will be in phase with the voltage across R, and I_C will lead V by $\pi/2$ radians. When the phasor diagram is drawn to scale it is seen that the resultant of I_C and I_R is 1·7 amperes. This agrees with the total current I as measured by the ammeter.

81

$$I = I_C + I_R$$
$$I^2 = I_C^2 + I_R^2$$

Thus, when applying Kirchhoff's first law to any point in an a.c. circuit we must combine currents by phasor addition.

Now, what about the power in the circuit? We already know that the power in an a.c. circuit is given by $V \times I \times \cos\phi$ where V and I are the r.m.s. values of the supply voltage and current and ϕ is the phase difference.

In Fig.3.2 we see,

$$\cos\phi = \frac{I_R}{I} = \frac{1 \cdot 1}{1 \cdot 7} = 0 \cdot 646$$

$\therefore$ power $= V \times I \times \cos\phi = 200 \times 1 \cdot 7 \times 0 \cdot 646 = 220$ watts

The power in the circuit is the power in the resistor only, since the power in a pure capacitor connected in an a.c. circuit is zero.

Power in the resistor $= V_R \times I_R = 200 \times 1 \cdot 1 = 220$ watts.

The power indicated by the wattmeter corresponds closely to the calculated power.

Parallel a.c., circuit containing resistance and inductive, reactance

Experiment 3(ii) *To investigate the currents and power in a parallel network comprising resistance and inductance connected across an alternating voltage supply*

A circuit similar to that used in the previous experiment was connected but the capacitor was replaced by an inductor.

The alternator field current was adjusted until the output was 200 volts at 50 Hz, and the currents in each branch of the circuit and the power were observed.

V (V)	I (A)	I_L (A)	I_R (A)	P (W)
200	1·5	0·93	1·1	227

Now let us sketch the phasor diagram for this circuit by commencing with the voltage phasor (Fig. 3.3).

The current I_R through the resistor will be in phase with V, and I_L through the inductor will lag behind V by $\pi/2$ radians. The resultant current supplied by the alternator is measured from the phasor diagram as 1·45 amperes.

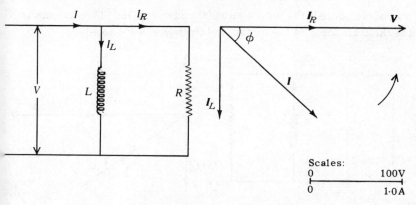

Fig. 3.3. Parallel $R-L$ circuit and phasor diagrams.

This agrees closely with the total supply current of 1·5 ampere as indicated by the ammeter.

$$I = I_L + I_R$$
$$I^2 = I_L^2 + I_R^2$$

Now, how can we calculate the power in the circuit?
In Fig. 3.3 we have,

$$\cos \phi = \frac{I_R}{I} = \frac{1\cdot 1}{1\cdot 45} = 0\cdot 76$$

Power = $V \times I \times \cos \phi$ = 200 × 1·5 × 0·76 = 228 watts.

Theoretically the power in a pure inductor is zero and hence the total power in the circuit will be the power in the resistor. Power in the resistor = $V_R \times I_R$ = 200 × 1·1 = 220 watts.

The power indicated by the wattmeter agrees reasonably closely with this calculated value. The difference is probably caused by a power loss in the coil L which is not quite purely inductive.

Parallel a.c., circuit containing resistance, inductance and capacitance

Experiment 3(iii) *To investigate the currents and power in a circuit network comprising resistance, inductance and capacitance connected in parallel*

A network consisting of the three components in parallel was connected across the output terminals of an alternator. The alternator field current was adjusted until the output was 200 volts at 50 Hz. The currents in the three branches of the circuit and the power were noted.

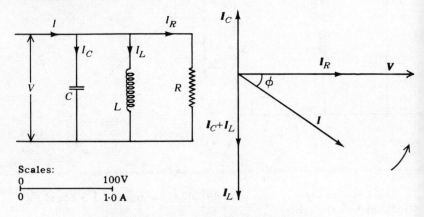

Fig. 3.4. Parallel L, C and R circuit and phasor diagram.

V (V)	I (A)	I_R (A)	I_C (A)	I_L (A)	P (W)
200	1·4	1·1	0·64	1·45	227

In drawing the phasor diagram we commence with the voltage phasor which is common to all components. The current I_R through the resistor is in phase with V while the current I_C through the capacitor leads V by $\pi/2$ radians. The resultant current supplied by the alternator is then given by,

$$I = I_R + I_L + I_C$$

As determined from the diagram the value of I is 1·4 ampere which agrees with the value indicated by the ammeter.

The total current in this parallel circuit is given by,

$$I = \sqrt{I_R^2 + (I_L - I_C)^2}$$

If the value of I_C is greater than I_L then the circuit is predominantly capacitive and the resulting power factor is leading but if I_L is greater than I_C then the inductive reactance predominates and the resulting power factor is lagging.

From the phasor diagram,

$$\cos \phi = \frac{I_R}{I} = \frac{1 \cdot 1}{1 \cdot 4} = 0 \cdot 785$$

and the power in the circuit,
$$V \times I \times \cos \phi = 200 \times 1{\cdot}4 \times 0{\cdot}785 = 220 \text{ watts}$$

This value again agrees reasonably closely with the wattmeter indication.

Summary

Impedance triangles may not be used when dealing with parallel connected components in a.c. circuits.

When resistance and capacitive reactance are connected in parallel,
$$I = \sqrt{I_R^2 + I_C^2}$$
$$\cos \phi = \frac{I_R}{I}$$

When resistance and inductive reactance are connected in parallel,
$$I = \sqrt{I_R^2 + I_L^2}$$
$$\cos \phi = \frac{I_R}{I}$$

In a parallel circuit with branches containing resistance, inductive reactance and capacitive reactance,
$$I = \sqrt{I_R^2 + (I_L - I_C)^2}$$

If $I_L > I_C$ the circuit is predominantly inductive and if $I_C > I_L$ the circuit is predominantly capacitive.

Examples 3.1.

1. An a.c. circuit consists of a 50 ohm resistor in parallel with a 40 μF capacitor connected across a 240 V 50 Hz supply.

Sketch the circuit and draw a phasor diagram representing the supply and branch currents. Calculate the total power and power factor.

2. A capacitor is shunted by a non-reactive resistor of 200 ohms and the combination takes a current of 2 A when connected to a 250 V 50 Hz supply. Calculate the capacitance and the power factor of the combination.

3. A choking coil of inductance 0·08 H and resistance 10 Ω is connected in parallel with a capacitor of 100 μF capacitance. The combination is connected to a supply of 200 V 50 Hz. Determine the total current taken from the supply and also the power factor. Illustrate your answer with a phasor diagram.

4. A capacitor of 51 µF is connected in parallel with a coil having a resistance of 25 ohms and a reactance of 30 ohms across a 200 V 50 Hz supply. Calculate the power factor of the current taken from the supply. What would be the power factor if the capacitance was doubled?

5. The following circuits A B and C are connected in parallel across a 240 volt 50 Hz supply:

A is a coil of resistance 5 ohms and 20 mH inductance.

B is a capacitor of 212 µF in series with a non-inductive resistor of 10 ohms.

C is a non-inductive resistor of 10 ohms.

Calculate for each circuit (a) the current (b) the power factor, and (c) the power dissipated. Determine by means of a phasor diagram drawn to scale the magnitude of the total supply current and the power factor of the combined circuits.

6. The load on a 250 volt supply system is:-

25 A at 0·6 power factor lagging,

15 A at 0·5 power factor lagging,

30 A at unity power factor,

and 16 A at 0·6 power factor leading.

Determine the total load in kVA and its power factor.

7. A capacitor of 1 µF is connected in series with a resistor of 1000 ohm across a sinusoidal a.c. supply of frequency 398 Hz. A resistor of 2000 ohms is also connected across the supply. If the voltage across the 1000 ohm resistor is 1 volt, determine by means of a phasor diagram the voltage of the supply, the current in the 2000 ohm resistor, the total current supplied and the phase difference between the supply current and voltage. What is the total power absorbed by the network?

8. A coil takes a current of 10 amperes when connected to a 200 volt 50 Hz supply, the power factor being 0·707. When cônnected in parallel with another coil the total current is 20 amperes and the resultant power factor is 0·866. Calculate the current when these coils are connected in series across the same supply.

The parallel resonant circuit

We already know how to deal with parallel circuits of resistance, inductance and capacitance connected across an alternating voltage supply.

Let us now have a look at such a parallel circuit when the frequency of the supply is varied.

Experiment 3 (iv) *To investigate the currents in the various branches of a parallel L–C circuit when the frequency of the supply is varied*

In this experiment a capacitor and inductor were connected in parallel across the output of an alternator driven by a variable speed motor. The magnitude of the voltage applied to the circuit was maintained at 80 volts throughout the test and the frequency varied over the range of 25 Hz to 80 Hz.

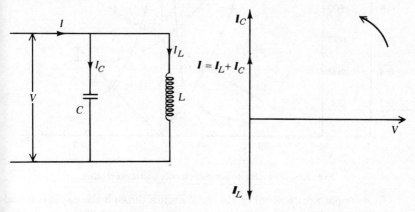

Fig. 3.5. Parallel $L-C$ circuit and phasor diagram.

The instrument readings were observed and tabulated

V	(V)	80	80	80	80	80	80	80
f	(Hz)	25	30	40	50	60	70	80
I	(A)	1·05	0·78	0·41	0·1	0·1	0·3	0·48
I_C	(A)	1·35	1·1	0·8	0·65	0·57	0·49	0·44
I_L	(A)	0·25	0·32	0·44	0·55	0·66	0·77	0·9
$Z = V/I$	(Ω)	76	103	195	800	800	266	167

In Fig. 3.6. these observations are plotted to a base of frequency

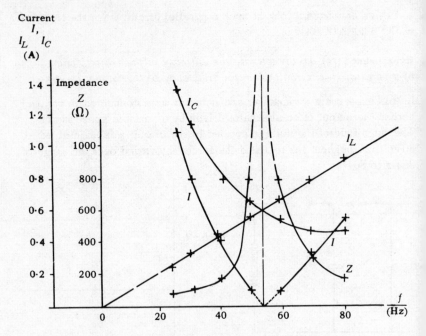

Fig. 3.6. Parallel resonant circuit characteristics.

At a frequency of about 55 Hz the currents through the capacitor and inductor branches of the circuit are equal in magnitude but have a phase displacement of π radians. The current from the supply is very low at this frequency and consequently the impedance of the parallel network is high.

The frequency at which the impedance is a maximum is termed the resonant frequency for the parallel circuit and the circuit may be referred to as a rejector circuit. At this frequency currents many times greater than the supply current may be circulating.

Neglecting the presence of any resistance in the inductive coil, resonance occurs when

$$X_L = X_C$$
$$\therefore 2\pi f L = \frac{1}{2\pi f C} \quad \text{and} \quad f = \frac{1}{2\pi \sqrt{LC}} \text{ Hz}$$

Parallel circuit current magnification

Some resistance is inevitably present in an inductive coil. Let us investigate the effect of this in a parallel resonant circuit.

Consider the circuit and corresponding phasor diagram in Fig.3.7.

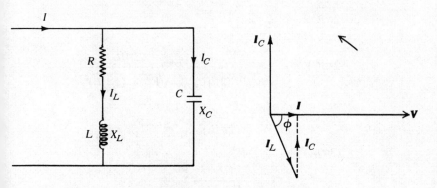

Fig. 3.7. Parallel resonant circuit.

The current through the inductive coil lags behind the voltage by an angle less than $\pi/2$ radians because of the coil resistance,

$$\cos \phi = \frac{R}{Z} \text{ where } Z \text{ is the impedance of the coil.}$$

$$\tan \phi = \frac{X_L}{R} = \frac{2\pi fL}{R}$$

The current magnification Q is the ratio

$$\frac{\text{capacitor current}}{\text{supply current}} = \frac{I_C}{I}$$

When the supply current and voltage are in phase,

$$Q = \frac{I_C}{I} = \tan \phi = \frac{2\pi fL}{R}$$

Hence the current magnification for a parallel circuit,

$$Q = \frac{2\pi fL}{R}$$

is the same expression as was obtained for the voltage magnification in a series circuit.

Resonance will occur when $I_C = I_L \sin \phi$

$$\text{or } 2\pi f C V = I_L \times \frac{X_L}{Z}$$

$$2\pi f C V = \frac{V}{\sqrt{R^2 + (2\pi f L)^2}} \times \frac{2\pi f L}{\sqrt{R^2 + (2\pi f L)^2}}$$

$$C = \frac{L}{R^2 + (2\pi f L)^2}$$

$$(2\pi f L)^2 = \frac{L}{C} - R^2$$

$$f = \frac{1}{2\pi L}\sqrt{\frac{L}{C} - R^2}$$

If R^2 is small with respect to L/C we have the expression for the resonant frequency as obtained earlier.

$$f = \frac{1}{2\pi\sqrt{LC}}$$

At resonance the network impedance $= \dfrac{V}{I}$ $\left(\tan \phi = \dfrac{I_C}{I}\right)$

$$= \frac{V}{I_C} \tan \phi$$

If R is small with respect to $2\pi f L$, $\tan \phi \simeq \dfrac{I_L}{I}$

$$\text{Circuit impedance} = \frac{1}{2\pi f C} \times \frac{2\pi f L}{R} = \frac{L}{CR} \text{ ohms}$$

This is termed the dynamic impedance of the circuit and the smaller we make R then the higher is the dynamic impedance.

In the parallel circuit the inductive and capacitive components constitute a closed circuit through which an interchange of energy may take place. At resonance the energy released by the discharging capacitor establishes a magnetic field linking with the inductor. This field in turn collapses and the energy stored is released to establish an electric field associated with the capacitor. Under ideal conditions the rates at which energy is released and stored are equal at resonance. Internal alternating currents circulate without the necessity for any input supply current and the circuit behaves as a rejector circuit with a very high impedance.

Summary

Resonance occurs in a parallel network of inductive and capacitive reactance when $f = 1/2\pi\sqrt{LC}$.

At this frequency the circuit presents a very high impedance to the supply and consequently the supply current is at a minimum.

When the inductive branch of the circuit includes resistance R this has an effect on the resonant frequency which is then given by

$$f = \frac{1}{2\pi L}\sqrt{\frac{L}{C} - R^2}$$

The circuit magnification at resonance,

$$Q = \frac{\text{capacitor current}}{\text{supply current}} = \frac{2\pi f L}{R}$$

The dynamic impedance given by L/CR is the impedance which the network presents to the supply at resonance.

Examples 3.2.

1. What do you understand by the term resonance in a parallel circuit? Calculate the resonant frequency of a parallel circuit consisting of a 200 µH inductor and a 450 pF capacitor.

2. An inductive coil takes a total current of 20 amperes and dissipates 3000 watts when connected to a 240 volt, 50 Hz supply. When a capacitor is connected in parallel with the coil across the same supply it is noted that the magnitude of the total current remains the same. Sketch a phasor diagram for the circuit and use it to explain the conditions existing. Calculate the value of the capacitor.

3. What is meant by resonance in (a) a series circuit and (b) a parallel circuit? Compare the impedances of the two circuits at resonance.

A coil of resistance R ohms and inductance L henrys is connected in parallel with a capacitor of C farads across a 100 volt variable-frequency supply. Sketch a phasor diagram representing the supply voltage and the circuit currents. Derive an expression for the frequency at which the circuit will behave as a pure resistor. If the coil has a resistance of 10 ohms and an inductance of 0·2 henry and C is 50 microfarads calculate the value of this frequency.

4. A sinusoidal alternating voltage at 1592 Hz is applied across the

terminals AB of the circuit in Fig.3.8. and the voltage across the 100 ohms resistor is found to be 4 volts. By means of a phasor diagram or otherwise determine: (a) the voltage of the supply (b) the total power supplied and (c) the power factor of the complete circuit. What value of capacitance connected between A and B would make the power factor of the complete circuit unity?

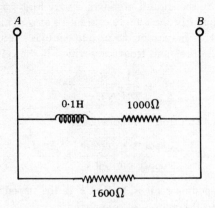

Fig. 3.8. Example 4.

Representing phasors by a symbolic notation

If you have worked through most of the examples set at the end of each section of this and the preceding chapters then you will have become familiar with phasor quantities. You will have carried out many calculations involving the squaring of quantities, the taking of square roots, resolving phasors into components and recombining components to produce phasors.

In the remainder of this chapter we are going to deal with a method of expressing phasors which will help us to simplify arithmetical working in future studies. We must not forget however that the complex equations which we shall use represent phasors, and that these in turn represent alternating quantities.

The symbolic notation is sometimes referred to as the j notation. A phasor is identified in this notation by quoting its horizontal and vertical components. To distinguish between these components the vertical component is preceded by the letter, or operator, j.

For instance in Fig.3.9. the phasor may be given in the form
$$\mathbf{V} = a + jb$$
Multiplying any quantity by j has the effect of rotating it through

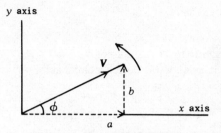

Fig.3.9. Phasor components.

$\pi/2$ radians (90°) in an anticlockwise direction. When dealing with phasor quantities the reference axis is normally taken as the x axis.

You will also observe in Fig.3.9. that the angle between the phasor **V** and the horizontal or x axis is given by $\phi = \tan^{-1} \frac{b}{a}$

If we wished to determine the magnitude of the phasor then this would be given by $V = \sqrt{a^2 + b^2}$

The magnitude of the phasor is sometimes termed the modulus.

Example

Determine the magnitude of the phasor **V** $= 3 + j4$, and the angle which it makes with the x axis.

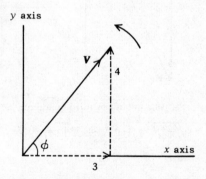

Fig.3.10. Components of a phasor.

Fig.3.10 shows the phasor,

$$\mathbf{V} = 3 + j4$$
$$V = \sqrt{3^2 + 4^2} = 5$$
$$\phi = \tan^{-1} \tfrac{4}{3} = 1\cdot 33 \quad \therefore \phi = 53\cdot 2°$$

This may be written as $V = 5 \angle 53\cdot 2°$

Here is a further example.

Example

Determine the magnitude of the phasor $\mathbf{V} = -3 - j6$.
What angle does it make with the horizontal axis?

$$\mathbf{V} = a + jb$$
$$\therefore \mathbf{V} = -3 - j6$$

This indicates that the component on the horizontal axis has a value of -3 and the component on the vertical axis has a magnitude of -6.

Figure 3.11 shows the position of the phasor which lies in the third quadrant.

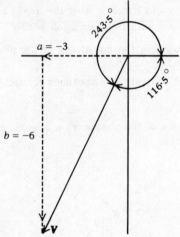

Fig. 3.11. Phasor in the third quadrant.

$$V = \sqrt{(-3)^2 + (-6)^2} = \sqrt{45} = 6\cdot 7$$
$$\phi = \tan^{-1}\frac{-6}{-3} = 243\cdot 5°$$
$$V = 6\cdot 7 \angle 243\cdot 5°$$

Alternatively this may be written $V = 6\cdot 7 \angle 116\cdot 5°$

Summary

A phasor may be expressed by stating the horizontal and vertical components.

$$\mathbf{V} = a + jb \quad \phi = \tan^{-1}\frac{b}{a}$$
$$V = \sqrt{a^2 + b^2}$$

Examples 3.3.

1. Calculate the magnitude and phase angles of phasors represented by the following complex equations and check your answers by drawing phasor diagrams to scale.

   (a) $V_1 = 4 + j8$
   (b) $V_2 = 3 + j4$
   (c) $V_3 = -7 - j8$
   (d) $V_4 = 6 - j12$
   (e) $V_5 = 10 - j4$
   (f) $V_6 = -8 + j2$

2. A current phasor has a magnitude of 12 amperes and makes an angle of 30° with the x axis. Express this phasor in the j notation.

3. A series circuit of resistance and capacitance takes a current of $I = 12 + j6$ from a 240 volt 50 Hz supply. Calculate the value of the capacitance.

Addition of phasors using the j notation

Consider two phasors $V_1 = a_1 + jb_1$ and $V_2 = a_2 + jb_2$. These two phasors and also the resultant phasor obtained by the addition of the individual phasors are shown in Fig. 3.12.

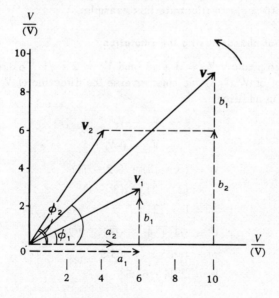

Fig. 3.12. Addition of phasors.

Clearly the resultant phasor has a horizontal component $a_1 + a_2$ and a vertical component $b_1 + b_2$

$$V_1 + V_2 = a_1 + a_2 + jb_1 + jb_2$$

Hence to add phasors expressed in the j notation we collect like terms and express in the form $a + jb$.

$$\text{Also, } \phi = \tan^{-1} \frac{b_1 + b_2}{a_1 + a_2} = \tan^{-1} \frac{b}{a}$$

Example

If $V_1 = 6 + j3$ and $V_2 = 4 + j6$ express the resultant phasor of addition in the j notation. Determine the magnitude and phase angle which the resultant makes with the x axis.

$$V_1 + V_2 = (6 + j3) + (4 + j6) = 10 + j9$$

The magnitude of the resultant phasor

$$V = \sqrt{100 + 81} = \sqrt{181} = 13\cdot 5 \, V$$

$$\phi = \tan^{-1} \tfrac{9}{10} = 42°$$

These results may be checked by referring to Fig.3.12. which has been drawn to scale to illustrate this example.

Subtraction of phasors using the j notation

Consider two phasors $V_1 = 3 + j3$ and $V_2 = 2 + j1$ To determine the resultant of $V_1 - V_2$ we must reverse the direction of V_2 and proceed as in addition.

$$V = V_1 - V_2$$
$$= V_1 + (-V_2)$$
$$V = (3 + j3) + (-2 - j1)$$
$$= (3 - 2) + j(3 - 1)$$
$$= 1 + j2$$
$$\therefore V = \sqrt{1 + 4} = \sqrt{5} = 2\cdot 24 \, V$$
$$\phi = \tan^{-1} 2 = 63\cdot 5°$$
$$V = 2\cdot 24 \angle 63\cdot 5°$$

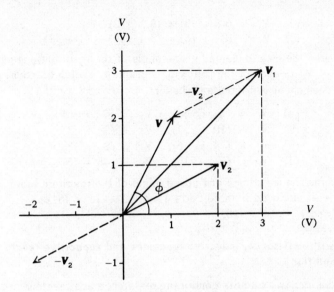

Fig.3.13. Substraction of phasors.

Summary

When combining phasors we combine horizontal and vertical components separately.

If, $\quad V_1 = a_1 + jb_1 \quad$ and $\quad V_2 = a_2 + jb_2$

then the sum, $\quad V = V_1 + V_2$

$\quad = a_1 + a_1 + jb_1 + jb_2$

and $\quad \phi = \tan^{-1} \dfrac{b_1 + b_2}{a_1 + a_2}$

Also the difference $V = V_1 - V_2$

$\quad = a_1 - a_2 + jb_1 - jb_2$

and $\quad \phi = \tan^{-1} \dfrac{b_1 - b_2}{a_1 - a_2}$

Examples 3.4.

1. Determine the resultant phasors of the following phasor additions. Sketch diagrams showing component and resultant phasors.

(a) $\quad V_a = (1 + j8) + (3 + j4)$
(b) $\quad V_b = (8 - j6) + (-7 + j3)$

(c) $\mathbf{V}_c = (-6 + j8) + (8 - j10)$
(d) $\mathbf{V}_d = (2 + j6) + (3 + j10) + (4 + j8)$

2. Determine the magnitude and phase angle of the resultant phasor $\mathbf{V} = \mathbf{V}_1 - \mathbf{V}_2$ in each of the following examples. Sketch diagrams showing component and resultant phasors.

(a) $\mathbf{V}_1 = (6 + j8)$, $\mathbf{V}_2 = (4 + j6)$
(b) $\mathbf{V}_1 = (10 + j7)$, $\mathbf{V}_2 = (3 + j8)$
(c) $\mathbf{V}_1 = (-4 + j6)$, $\mathbf{V}_2 = (8 + j8)$
(d) $\mathbf{V}_1 = (6 - j10)$, $\mathbf{V}_2 = (-8 - j6)$

3. Determine the magnitude and phase angle of the resultant e.m.f. $e_a - e_b + e_c$ induced in three coils when $e_a = 12 + j6$, $e_b = 10 + j4$ and $e_c = 4 + j10$.

Dealing with resistance, inductive reactance and capacitive reactance in the j notation

Let us consider two circuits containing resistance and reactance.

(a) *R–L circuit.*

We have seen earlier in Chapter 2 how the impedance triangle may be derived from the $R-L$ series circuit phasor diagram.

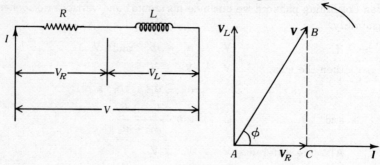

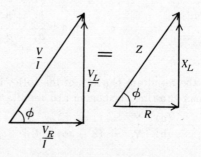

Fig. 3.14. $R-L$ impedance triangle.

The relationship expressed by the impedance triangle may be written in the form

$$Z = R + X_L$$
$$\therefore Z = R + j\omega L$$

Thus if resistance is represented by R then inductive reactance is represented by $j\omega L$

(b) *R–C circuit*

You have also seen earlier how the impedance triangle may be derived from the $R-C$ series circuit phasor diagram.

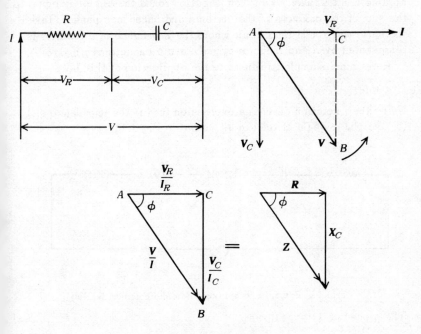

Fig. 3.15. $R-C$ impedance triangle.

The relationship expressed by the impedance triangle may be written in the form

$$Z = R + X_C$$
$$\therefore Z = R - j\frac{1}{\omega C}$$
$$= R + j\left(-\frac{1}{\omega C}\right)$$

Thus if resistance is represented by R then capacitive reactance is represented by $j(-1/\omega C)$.

Multiplying a phasor by $-j$ has the effect of turning it through $\pi/2$ radians (90°) in a clockwise direction.

The results of our investigation into these two impedance triangles enables us to apply Kirchhoff's second law in an amended form to a.c. circuits.

How may Kirchhoff's second law be applied to a.c. circuits?

The amended law may be stated as follows:

For any closed circuit the sum of the phasors representing the applied e.m.f's taken in a given direction round the circuit is equal to the sum of the products of the currents and impedance phasors taken the same way round the circuit where the impedance of a resistor is represented by R, an inductor by $j\omega L$, and a capacitor by $j(-1/\omega C)$.

Here is an example to illustrate the application of this law.

Example

Using the j notation derive an expression for (a) the impedance and (b) the phase angle of the circuit shown in Fig.3.16.

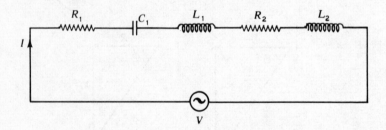

Fig.3.16. Example on applying Kirchhoff's amended law.

The procedure is as follows:

1. Draw the circuit diagram
2. Apply Kirchhoff's second law to determine $Z = \dfrac{V}{I}$
3. Express the impedance in the form $Z = a + jb$

$$\text{when } \phi = \tan^{-1} \frac{b}{a}$$

$$V = IR_1 + Ij\left(-\frac{1}{\omega C}\right) + Ij\omega L_1 + IR_2 + Ij\omega L_2$$

$$\therefore \mathbf{Z} = \frac{V}{I} = R_1 + j\left(-\frac{1}{\omega C}\right) + j\omega L_1 + R_2 + j\omega L_2$$

$$\mathbf{Z} = (R_1 + R_2) + j\left(\omega L_1 + \omega L_2 - \frac{1}{\omega C}\right)$$

$$\text{and } \mathbf{Z} = \sqrt{(R_1 + R_2)^2 + \left(\omega L_1 + \omega L_2 - \frac{1}{\omega C}\right)^2}$$

$$\phi = \tan^{-1}\left(\frac{\omega L_1 + \omega L_2 - \dfrac{1}{\omega C}}{R_1 + R_2}\right)$$

Here is a further example with numerical values.

Example

Determine the total impedance and phase angle for the circuit shown in Fig.3.17 when the supply frequency is 50 Hz

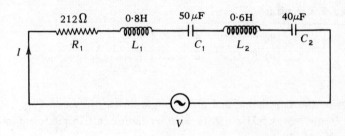

Fig.3.17. Series a.c. circuit.

$$\mathbf{V} = IR_1 + Ij\omega L_1 + Ij\left(-\frac{1}{\omega C_1}\right) + Ij\omega L_2 + Ij\left(-\frac{1}{\omega C_2}\right)$$

$$\mathbf{Z} = \frac{V}{I} = R_1 + j\omega L_1 + j\left(-\frac{1}{\omega C_1}\right) + j\omega L_2 + j\left(-\frac{1}{\omega C_2}\right)$$

$$= R_1 + j\left(\omega L_1 + \omega L_2 - \frac{1}{\omega C_1} - \frac{1}{\omega C_2}\right)$$

$$\mathbf{Z} = \sqrt{R_1^2 + \left(\omega L_1 + \omega L_2 - \frac{1}{\omega C_1} - \frac{1}{\omega C_2}\right)^2}$$

$$= \sqrt{212^2 + \left[(314 \times 0 \cdot 8) + (314 \times 0 \cdot 6) - \frac{10^6}{314 \times 50} - \frac{10^6}{314 \times 40}\right]^2}$$

$$= \sqrt{212^2 + (251 + 188 - 63 \cdot 6 - 79 \cdot 6)^2}$$

$$= \sqrt{212^2 + 296^2}$$

$$Z = \sqrt{45\,000 + 88\,000} = \sqrt{133\,000}$$
$$= 364\,\Omega$$
$$\phi = \tan^{-1}\frac{296}{212} = \tan^{-1} 1\cdot 4 = 54\cdot 4°$$
$$Z = 364 \angle 54\cdot 4°$$

Since the numerical value 296 is positive this indicates that $X_L > X_C$ and the circuit is predominantly inductive. Had the value of 296 been negative then this would have meant that $X_C > X_L$ and the circuit would have been presominantly capacitive.

Summary

Kirchhoff's amended law requires,
$$X_L = j\omega L \text{ and } X_C = j\left(-\frac{1}{\omega C}\right)$$

In series a.c. circuits,

(a) R–L:- $\mathbf{Z} = R + j\omega L$
(b) R–C:- $\mathbf{Z} = R + j\left(-\dfrac{1}{\omega C}\right)$

Examples 3.5.

1. Determine the impedance and phase angle of the circuit in Fig.3.18. when the frequency is 50 Hz. State whether the circuit is predominantly inductive or capacitive.

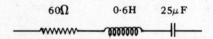

Fig.3.18. Example 1.

2. Determine the current and power factor for the circuit (Fig.3.19).

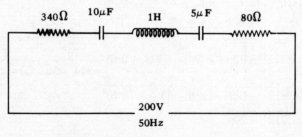

Fig.3.19. Example 2.

3. Determine the p.d. across the 100 ohm resistor in the circuit of Fig.3.20.

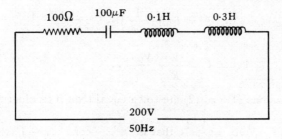

Fig.3.20. Example 3.

What happens if we meet j^2?

Consider a parallel circuit as in Fig.3.21.

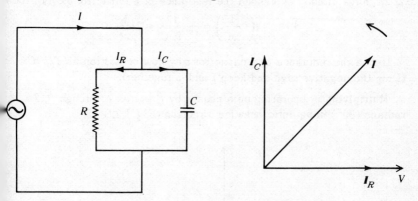

Fig.3.21. Parallel circuit and phasor diagram.

From the phasor diagram we see that

$$\begin{aligned}I &= I_R + jI_C \\ &= \frac{V}{R} + j\frac{V}{X_C} \\ &= \frac{V}{R} + jV\omega C \ldots \end{aligned} \qquad (1)$$

Now let us consider the two parallel paths and apply Kirchhoff's amended law. We may write,

$$V = I_R R \quad \therefore I_R = \frac{V}{R}$$

$$V = I_C \, j\left(-\frac{1}{\omega C}\right) \therefore I_C = \frac{V\omega C}{-j}$$

hence $I = \dfrac{V}{R} + \dfrac{V\omega C}{-j} \ldots$ \hfill (2)

If expressions (1) and (2) are to be equal then it is clear that,

$$j\,V\omega C = \frac{V\omega C}{-j} \text{ that is } j = \frac{1}{-j} \text{ or } j^2 = -1$$

Thus if in future we meet j^2 then we may substitute -1 for j^2.

Note

1. We have already expressed the reactance of a capacitor as $j(1/-\omega C)$

Now, $\quad j\left(-\dfrac{1}{\omega C}\right) = -\dfrac{j^2}{j\omega C} = \dfrac{1}{j\omega C}$

Hence the reactance of a capacitor may now be written as $1/j\omega C$ (Drop the negative sign and keep j and ω together).

2. Multiplying or operating on a phasor by j rotates it through $\pi/2$ radians (90°) in an anticlockwise direction (Fig.3.22)

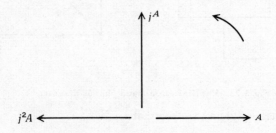

Fig.3.22. Multiplying by j and j^2.

If we multiply again by j the phasor is rotated through a further 90°. The effect of multiplying by j^2 is thus to reverse the direction of a phasor.

$$j^2 A = (-1) A = -A$$

What would be the effect of multiplying by j^3?

Multiplying and dividing phasor expressions

In problems which follow you will find it necessary to multiply and divide phasors which are expressed in the j notation.

Products

The product of two phasors is a phasor of length equal to the product of the individual phasor lengths. The resultant phase angle is equal to the sum of the individual phase angles. Consider this example.

Example

Determine the magnitude and direction of the phasor, $\mathbf{V} = (3 + j2)(7 + j3)$.

$$\begin{aligned}
\mathbf{V} &= (3 + j2)(7 + j3) \\
&= 21 + j9 + j14 + j^2 6 \\
&= (21 - 6) + j(9 + 14) = 15 + j23
\end{aligned}$$

$$V = \sqrt{15^2 + 23^2} = \sqrt{225 + 529} = \sqrt{754} = 27 \cdot 5 \text{ volts}$$

$$\phi = \tan^{-1} \frac{23}{15} = \tan^{-1} 1 \cdot 535 = 56 \cdot 9°$$

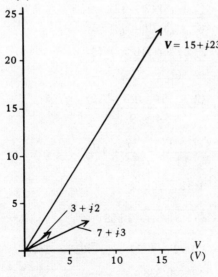

Fig. 3.23. Multiplying phasors.

The individual phasors are shown in Fig.3.23. and they make angles of

$$\phi_1 = \tan^{-1}\frac{2}{3} = 33\cdot 7° \text{ and } \phi_2 = \tan^{-1}\frac{3}{7} = 23\cdot 2°$$

with the horizontal.

It will be observed that the angle of the resultant phasor product is calculated to be $56\cdot 9°$ which is $33\cdot 7° + 23\cdot 2°$.

Quotients

The quotient of two phasors is a phasor of length equal to the quotient of the individual phasor lengths. The phase angle is the difference of the individual phase angles. Consider this example.

Example

Determine the magnitude and direction of the phasor

$$V = \frac{1 + j5}{2 + j3}$$

You will observe that we have a complex quantity, that is a term involving j, in the denominator. We must eliminate j from the denominator and to do this we multiply the numerator and denominator by $2 - j3$. This is known as rationalising.

$$V = \frac{(1 + j5)(2 - j3)}{(2 + j3)(2 - j3)}$$

$$= \frac{2 - j3 + j10 - j^2 15}{(2)^2 - (j3)^2}$$

$$= \frac{(2 + 15) - j(3 - 10)}{4 + 9}$$

$$= \frac{17 + j7}{13} = \frac{17}{13} + j\frac{7}{13}$$

$$= 1\cdot 31 + j0\cdot 538$$

$$V = \sqrt{(1\cdot 31)^2 + (0\cdot 538)^2}$$

$$= \sqrt{1\cdot 72 + 0\cdot 29} = \sqrt{2\cdot 01} = 1\cdot 42$$

$$\phi = \tan^{-1}\frac{0\cdot 538}{1\cdot 31} = \tan^{-1} 0\cdot 41 = 22\cdot 3°$$

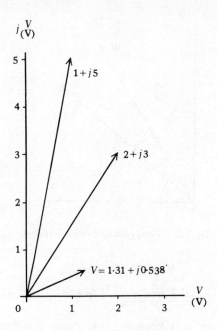

Fig. 3.24. Quotient of two phasors.

The individual phasors make angles of

$$\phi_1 = \tan^{-1} \frac{5}{1} = 78 \cdot 7°$$

and

$$\phi_2 = \tan^{-1} \frac{3}{2} = 56 \cdot 4°$$

with the horizontal axis.

It will be observed that the angle of the resultant quotient phasor as calculated is $22 \cdot 3°$ which is equal to the difference between $78 \cdot 7°$ and $56 \cdot 4°$

Note

To rationalize a quotient express the denominator in the form $(a + jb)$ and then multiply numerator and denominator by $(a - jb)$. Alternatively express the denominator in the form $(a - jb)$ and multiply the numerator and denominator by $(a + jb)$. This will cause j to vanish from the denominator.

Power expressed in the *j* notation

Consider two phasors **V** and **I** which make angles of ϕ_1 and ϕ_2 with a reference axis as in Fig. 3.25.

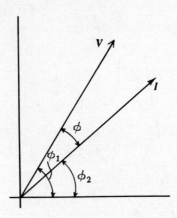

Fig. 3.25. Voltage and current phasors.

Let,
$$\mathbf{V} = a + jb = V \cos \phi_1 + V \sin \phi_1$$
and
$$\mathbf{I} = c + jd = I \cos \phi_2 + I \sin \phi_2$$

$$\begin{aligned}
\text{Power} &= VI \cos \phi \\
&= VI \cos (\phi_1 - \phi_2) \\
&= VI \cos \phi_1 \cos \phi_2 + VI \sin \phi_1 \sin \phi_2 \\
&= (ac) + (bd)
\end{aligned}$$

Hence the power is given by the sum of the products of the horizontal and vertical components.

$$\begin{aligned}
\text{Reactive power (VAr)} &= VI \sin \phi \\
&= VI \sin (\phi_1 - \phi_2) \\
&= VI \sin \phi_1 \cos \phi_2 - VI \cos \phi_1 \sin \phi_2 \\
&= (bc) - (ad)
\end{aligned}$$

Example

A voltage $\mathbf{V} = 6 + j8$ is applied to a circuit and the resulting current is $\mathbf{I} = 4 + j7$. Determine the power in the circuit and also the reactive voltamperes. Evaluate the supply voltamperes and show on a diagram how these are related to the watts and reactive voltamperes.

Let
$$\mathbf{V} = a + jb$$
$$\mathbf{I} = c + jd$$

$$\text{Power} = ac + bd$$
$$= (6 \times 4) + (8 \times 7) = 80\,\text{W}$$
$$\text{Reactive VA} = (bc) - (ad)$$
$$= (8 \times 4) - (6 \times 7) = -10\,\text{VA}$$

The negative sign indicates that the current leads the voltage.
Supply voltamperes,

$$\mathbf{V} \times \mathbf{I} = (6 + j8)(4 + j7)$$
$$= 24 + j42 + j32 + j^2 56$$
$$= 24 - 56 + j74$$
$$= -32 + j74$$
$$V \times I = \sqrt{(-32)^2 + (74)^2} = \sqrt{6500} = 80\cdot6\,\text{VA}.$$

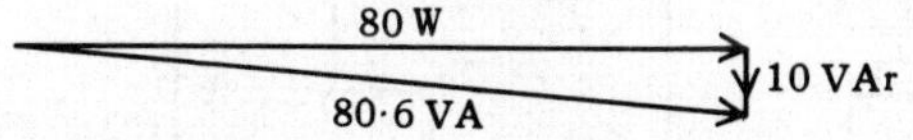

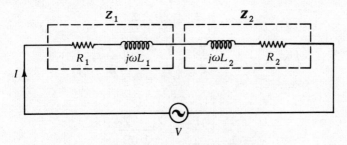

Fig.3.26. Relation between watts voltamperes and reactive voltamperes.

The diagram (Fig.3.26) shows how the watts, voltamperes and reactive voltamperes are related.

Impedances in series

Fig.3.27. Impedances in series.

To combine two series impedances $\mathbf{Z}_1$ and $\mathbf{Z}_2$ (Fig.3.27) expressed as phasor quantities we may apply Kirchhoff's amended second law.

$$V = IR_1 + Ij\omega L_1 + IR_2 + Ij\omega L_2$$

$$Z = \frac{V}{I} = (R_1 + j\omega L_1) + (R_2 + j\omega L_2)$$

$$Z = Z_1 + Z_2$$

Thus impedances in series may be combined arithmetically provided that the impedances are expressed as phasors in symbolic notation. The resultant impedance will be expressed in the same form.

Impedances in parallel

Consider two impedances Z_1 and Z_2 in parallel (Fig.3.28) Z_1 and Z_2 are phasors and may be combinations of resistance and reactance. For example, Z_1 may be comprised of $R + j\omega L_1$ while Z_2 may have $R_2 + 1/j\omega C$ components.

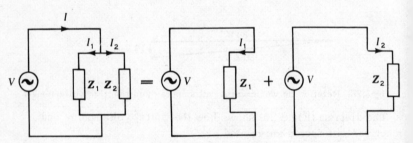

Fig.3.28. Impedances in parallel.

$$I_1 = \frac{V}{Z_1} \quad I_2 = \frac{V}{Z_2}$$

$$I = I_1 + I_2 = \frac{V}{Z_1} + \frac{V}{Z_2} = V\left(\frac{1}{Z_1} + \frac{1}{Z_2}\right)$$

$$\therefore Z = \frac{V}{I} = \frac{1}{\dfrac{1}{Z_1} + \dfrac{1}{Z_2}} = \frac{1}{\dfrac{Z_2 + Z_1}{Z_1 Z_2}}$$

$$Z = \frac{Z_1 Z_2}{Z_1 + Z_2} = \frac{\text{product of impedances}}{\text{sum of impedances}}$$

Thus impedances in two parallel a.c. paths may be combined using an expression similar to that for two resistors in parallel providing that the impedances are expressed in symbolic notation. The resultant impedance will be given in a similar symbolic form.

The branched circuit rule

Consider two impedances Z_1 and Z_2 across an alternating voltage supply (Fig.3.29).

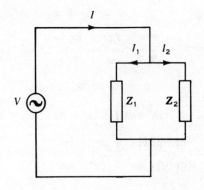

Fig.3.29. Branched circuit currents.

$$I_1 = \frac{V}{Z_1} \quad I_2 = \frac{V}{Z_2}$$

$$Z = \frac{V}{I} = \frac{Z_1 Z_2}{Z_1 + Z_2}$$

$$Z = \frac{Z_1 I_1}{I} = \frac{Z_1 Z_2}{Z_1 + Z_2}$$

$$\therefore I_1 = \frac{I Z_1 Z_2}{Z_1 (Z_1 + Z_2)} = I \frac{Z_2}{Z_1 + Z_2}$$

This equation may be stated in words. The phasor for the current through one branch of the circuit is the product of the phasor for the total current, and a phasor which is the ratio of the impedance of the other branch of the circuit to the phasor sum of the impedances in both branches.

Summary

$$j^2 = -1$$

$$V_1 = a_1 + jb_1 \qquad \phi_1 = \tan^{-1}\frac{b_1}{a_1}$$

$$V_2 = a_2 + jb_2 \qquad \phi_2 = \tan^{-1}\frac{b_2}{a_2}$$

Product of two phasors: $V_1 \times V_2 = (a_1 + jb_1)(a_2 + jb_2)$

$$\phi = \phi_1 + \phi_2$$

Quotient of two phasors: $\dfrac{V_1}{V_2} = \dfrac{(a_1 + jb_1)}{(a_2 + jb_2)}$

$$\phi = \phi_1 - \phi_2$$

Power   If $V = a + jb$ and $I = c + jd$

then power $= ac + bd$

and reactive voltamperes $= bc - ad$

Two impedances in series:

$$Z = Z_1 + Z_2$$

Two impedances in parallel

$$Z = \frac{Z_1 Z_2}{Z_1 + Z_2}$$

Branched circuit rule:

$$I_1 = I\frac{Z_2}{Z_1 + Z_2}$$

Examples 3.6.

1. What is the significance of applying the operators j and j^2 to a component of a current phasor? By referring to a parallel R–C circuit show that $j^2 = -1$.

2. Determine the phase angles of the following phasors with reference to the horizontal axis, $V_1 = 4 + j8$, $V_2 = 6 + j3$.

Calculate the product of the two phasors and show that the resultant phase angle is equal to the sum of the individual phase angles.

3. Determine the product of $(8 + j4)$, $(6 + j3)$ and $(3 + j4)$. Show that the resultant phase angle is equal to the sum of the individual phase angles.

4. What is meant by rationalising the denominator of a fraction which contains complex quantities?

Determine the magnitude and phase angle of the following phasor quotient,

$$V = \frac{6 + j4}{3 - j10}$$

Show that the resultant phase angle is equal to the difference of the individual phasor angles.

5. Determine the magnitude and phase angle of the impedance,

$$Z = \frac{(4 + j6)(7 + j9)}{(3 + j4)}$$

6. Write down expressions for the total impedance of the circuits shown in Fig.3.30

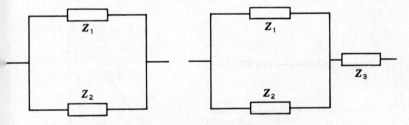

Fig.3.30. Impedance networks.

7. Determine the total impedance and the equivalent series resistance and reactance to the circuit of Fig.3.31. Values are in ohms.

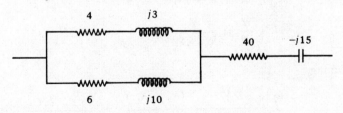

Fig.3.31. Examples 7 and 8.

8. Using the j notation calculate the power and reactive voltamperes in (a) the parallel section and (b) the complete circuit of Fig.3.31. when a 200 V a.c. supply is connected across the circuit.

Admittance, conductance and susceptance

In d.c. circuits the only quantity measured in ohms is the resistance. Conductance which is the reciprocal of resistance is measured in siemens.

In a.c. circuits however resistance, reactance and impedance are all measured in ohms and this means that reciprocals of combinations of these three quantities have the units of siemens.

The *Admittance* (Y) of an a.c. circuit is the reciprocal of impedance. Admittance may be looked upon as the current resulting when unit voltage is applied to a circuit.

$$Y = \frac{I}{V} = \frac{1}{Z}$$

(a) Circuit containing resistance and inductance

Consider a circuit as shown in Fig.3.32

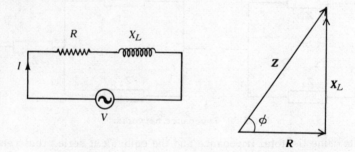

Fig.3.32. Series $R-L$ circuit with impedance triangles.

$$\mathbf{Z} = R + jX_L$$

and $$\mathbf{I} = \frac{\mathbf{V}}{\mathbf{Z}} = \frac{\mathbf{V}}{R + jX_L}$$

Rationalising,

$$\mathbf{I} = \frac{\mathbf{V}}{(R + jX_L)} \frac{(R - jX_L)}{(R - jX_L)} = \frac{\mathbf{V}(R - jX_L)}{R^2 + X_L^2}$$

$$\therefore Y = \frac{1}{Z} = \frac{I}{V} = \frac{R}{R^2 + X_L^2} - j\frac{X_L}{R^2 + X_L^2}$$

This expression for **Y** may be represented by a triangle which is known as the admittance triangle (Fig.3.33).

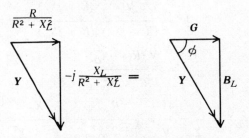

Fig.3.33. **G–B**$_L$ admittance triangle.

The admittance phasor (**Y**) thus has two components, the *Conductance* (**G**) and the *Inductive Susceptance* (**B**$_L$). Both are measured in siemens.

$$Y = G - jB_L$$

$$\tan\phi = \frac{B_L}{G}$$

It is important to note that **G** and **B**$_L$ are not simply the reciprocals of **R** and **X**$_L$ unless one of these resistance or reactance components in a circuit is zero.

(*b*) *Circuit containing resistance and capacitance.*

Consider a circuit as shown in Fig.3.34

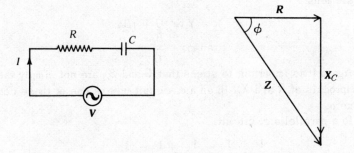

Fig.3.34. Series $R-C$ circuit with impedance triangle.

115

and
$$Z = R - jX_C$$
$$I = \frac{V}{Z} = \frac{V}{R - jX_C}$$

Rationalising,
$$I = \frac{V}{(R - jX_C)} \frac{(R + jX_C)}{(R + jX_C)}$$
$$= \frac{V(R + jX_C)}{R^2 + X_C^2}$$
$$\therefore Y = \frac{1}{Z} = \frac{I}{V} = \frac{R}{R^2 + X_C^2} + j\frac{X_C}{R^2 + X_C^2}$$

This expression for **Y** may be represented by an admittance triangle (Fig. 3.35).

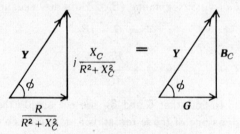

Fig. 3.35. **G–B** admittance triangle.

The admittance phasor (**Y**) thus has two components, the *Conductance* (**G**) and the *Capacitive Susceptance* (**B**$_C$). Both components are measured in siemens.

$$Y = G + jB_C$$
$$\tan \phi = \frac{B_C}{G}$$

Again it is important to stress that **G** and **B**$_C$ are not simply the reciprocals of **R** and **X**$_C$ in an a.c. circuit unless one of these components is zero.

In a parallel a.c. circuit,
$$\frac{1}{Z} = \frac{1}{Z_1} + \frac{1}{Z_2} + \frac{1}{Z_3} + \frac{1}{Z_4} + \ldots$$
$$Y = Y_1 + Y_2 + Y_3 + Y_4 + \ldots$$

It is sometimes easier to work in terms of admittance rather than impedance when dealing with parallel circuits.

Here is an example.

Example

Determine the total supply current and the phase angle for the network shown in Fig.3.36.

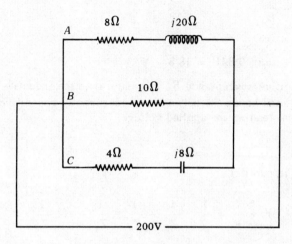

Fig.3.36. Example.

$R-L$ series circuit

$$\frac{R}{R^2 + X_L^2} - j\frac{X_L}{R^2 + X_L^2} = G - jB_L$$

$R-C$ series circuit

$$\frac{R}{R^2 + X_C^2} + j\frac{X_C}{R^2 + X_C^2} = G + jB_C$$

Branch A
$G = \dfrac{8}{8^2 + 20^2} = 0\cdot017\,S$ $B_L = -\dfrac{20}{8^2 + 20^2} = -0\cdot043\,S$

Branch B
$G = \dfrac{10}{10^2} = 0\cdot100\,S$ $B = 0\cdot0$

Branch C
$G = \dfrac{4}{4^2 + 8^2} = 0\cdot05\,S$ $B_C = \dfrac{8}{4^2 + 8^2} = 0\cdot1\,S$

Total $G = 0\cdot167\,S$ $B_C = 0\cdot057\,S$

$$Y = \sqrt{0\cdot 167^2 + 0\cdot 057^2} = \sqrt{0\cdot 0278 + 0\cdot 0033} = \sqrt{0\cdot 031}$$
$$= 0\cdot 176\, S$$
$$I = V \times Y$$
$$= 200 \times 0\cdot 176$$
$$= 35\cdot 2\ A$$
$$\tan \phi = \frac{B}{G}$$
$$\therefore \phi = \tan^{-1} 0\cdot 341 = 18\cdot 8°$$

The capacitive susceptance B_C is greater than the inductive susceptance B_L and hence the network is predominantly capacitive — the supply current leading the applied voltage.

Summary

Impedances in parallel,

Total *Admittance*,
$$\mathbf{Y} = \frac{I}{\mathbf{V}} = \frac{1}{\mathbf{Z}} = \frac{1}{\mathbf{Z}_1} + \frac{1}{\mathbf{Z}_2} + \frac{1}{\mathbf{Z}_3} + \ldots \frac{1}{\mathbf{Z}_n}$$

$R-L$ in series
$$\mathbf{Y} = \frac{R}{R^2 + X_L^2} - j\frac{X_L}{R^2 + X_L^2}$$
$$\mathbf{Y} = G - jB_L$$

$R-C$ in series
$$\mathbf{Y} = \frac{R}{R^2 + X_C^2} + j\frac{X_C}{R^2 + X_C^2}$$
$$\mathbf{Y} = G + jB_C$$

Conductances in parallel
$$G = G_1 + G_2 + G_3 + \ldots + G_n$$

Susceptances in parallel
$$B = B_1 + B_2 + B_3 + \ldots + B_n$$

Examples 3.7.

1. A series circuit comprises a resistor of 30 ohms in series with an inductive reactance of 25 ohms across an a.c. supply. Determine the impedance, admittance, conductance, and inductive susceptance of the circuit. What is the power factor?

2. A 200 µF capacitor and 30 ohm resistor are connected in series across a 100 volt 50 Hz supply. Determine the conductance, susceptance and admittance of the circuit. What effect would doubling the frequency of the applied voltage have on these quantities and what would then be the current taken from the supply?

3. A parallel network consists of two impedances represented by $5 + j6 \cdot 28$ and $2 + j1 \cdot 6$ respectively. Determine the total admittance and current when an alternating p.d. of 40 volts is applied. What is the power factor of the circuit?

4. Determine the total admittance, conductance, impedance and susceptance of a parallel a.c. network comprising two branches with impedances represented by $Z_A = 8 + j7$ and $Z_B = 12 - j7$. What would be the resistance, reactance and impedance of an equivalent series circuit?

5. Determine the total equivalent series resistance and reactance to (a) the parallel section of the network and (b) to the complete network shown in Fig.3.37.

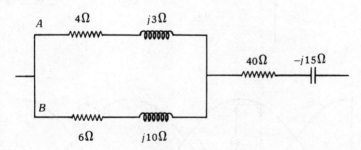

Fig. 3.37. Network of Example 5.

4. Three-Phase Connections, the Alternator, and Power in Three-Phase Systems

Introduction

In the alternating current experiments already carried out we have been using only one of the three windings on the alternator rotor. We now come to the study of three-phase systems where all three windings on the rotor are brought into use.

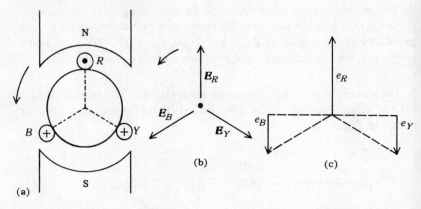

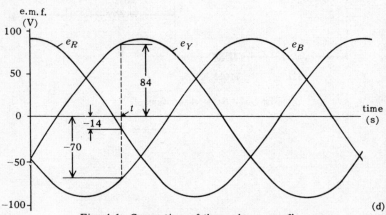

Fig. 4.1. Generation of three-phase e.m.f's.

Generation of three-phase e.m.f's

Consider an armature rotating in a magnetic field (Fig. 4.1(a)). The armature carries three equally spaced conductors marked R (red), Y (yellow) and B (blue).

The e.m.f's induced in the conductors will be $2\pi/3$ radians out of phase with each other and they may be represented by three rotating phasors (Fig. 4.1(b)). The developed waves are shown in Fig. 4.1(d). The phase sequense is the order in which the e.m.f's reach their maximum positive values and in this case it is red, yellow and blue.

Figure 4.1(c) shows the instantaneous values of the e.m.f's induced in the conductors at the instant for which Fig. 4.1(a) is drawn. It will be observed that at any instant the sum of the induced e.m.f's is zero. For example at the instant t,

$$e_R + e_Y + e_B = (-14) + 84 + (-70)$$
$$= 0 \text{ volts}$$

Let us now consider three single-turn coils on a rotor, that is the rotating part of a machine as in Fig. 4.2.

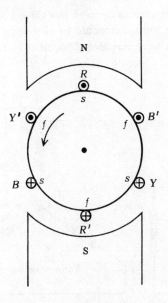

Fig. 4.2. Three single turn coils.

As the rotor is rotated the e.m.f's induced in both sides of each coil assist one another. By providing slip rings to which the ends of the coils are connected, three entirely independent outputs may be obtained (Fig. 4.3). Carbon brushes which are held by springs against each slip ring enable connections to be made with external circuits.

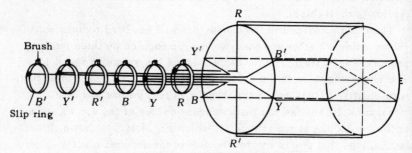

Fig. 4.3. Connections from coils to slip rings.

How the line and phase voltages in a star connected alternator vary with increasing field current

Experiment 4(i) *To investigate the e.m.f's induced in the windings of a three-phase star-connected alternator as the field current is increased*

In this experiment the alternator used has three rotor windings. The ends of each winding were connected to slip rings and the 'starts' (s) of the three windings were connected together to form a star or neutral point N (Fig. 4.4).

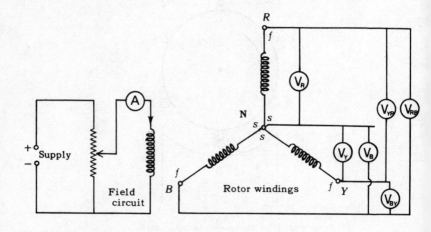

Fig. 4.4. Three-phase star-connected alternator.

The alternator was driven on no load by a d.c. motor and the motor speed was adjusted to produce an alternator output at 50 Hz. The alternator field current was increased in steps and the e.m.f's induced in individual windings and between the finishes (s) of different windings were observed and tabulated. Since the alternator was on no load the voltmeter readings were assumed to indicate induced e.m.f.'s.

Field current	(A)	0	0·5	1·0	1·5	2·0	2·5	3·0
E_R	(V)	0	18	37	61	80	100	118
E_Y	(V)	0	18	37	61	80	100	118
E_B	(V)	0	18	37	61	80	100	118
E_{YR}	(V)	0	30	63	105	138	173	204
E_{RB}	(V)	0	30	63	105	138	173	204
E_{BY}	(V)	0	30	63	105	138	173	204

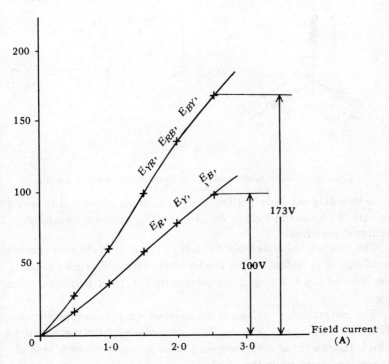

Fig. 4.5. Three-phase star-connected alternator voltage characteristics.

It is seen from the results that the e.m.f. induced between two of the terminals R, Y and B is greater than that induced in a single coil. For example, with a field current of 2·5 amperes the e.m.f. induced between any two finishes is 173 volts while that induced between the start and finish of one coil is only 100 volts.

In general the e.m.f. induced between two of the terminals R, Y and $B = 1·73 \times$ e.m.f. induced in any one coil.

How can we account for this?

Using a phasor diagram to determine the relationship between the line and phase voltages of a star-connected alternator

With the 'starts' of the three coils connected together to form a neutral or star point N we may represent the winding by the diagram of Figure 4.6(a).

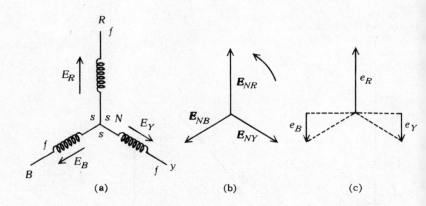

Fig. 4.6. Phasor representation of three-phase induced e.m.f's.

In this diagram E_R signifies 'the r.m.s. value of the e.m.f. induced in coil R'. Arrows alongside E_R, E_Y and E_B indicate directions assumed positive.

The diagram (b) represents the e.m.f's induced in the star connected windings. E_{NR} refers to 'the phasor representing the r.m.s. value of the induced e.m.f. acting in a direction from N to R within the winding'.

The instantaneous values of the induced e.m.f's are represented in diagram (c) by e_R, e_Y and e_B. The induced e.m.f's at the instant for which diagram (b) is drawn are seen to be negative in the Y and B phases and positive in the R phase.

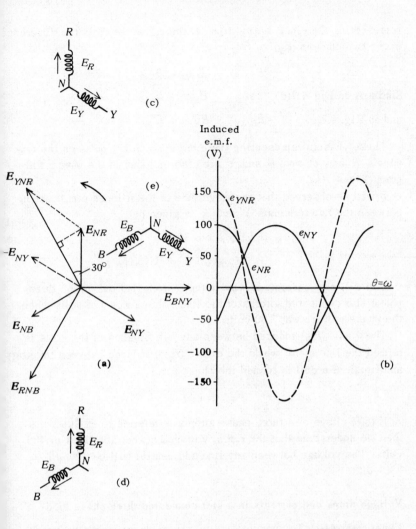

Fig. 4.7. Phasor representation of line and phase induced e.m.f's.

The diagrams (Fig. 4.7(a & b)) are phasor and wave diagrams representing the e.m.f's induced in the windings. The instantaneous value of the e.m.f. e_{YNR} (Fig. 4.7(c)) induced from Y through N to R is obtained by combining the ordinates of wave e_{NY} with those of e_{NR}. The resulting wave e_{YNR} is seen to be a sine wave with a value when $t = 0$ of $e_{NR} - e_{NY} = 100 - (-50) = 150$ volts. The instantaneous value of e_{YNR} is decreasing and hence the phasor representing the resulting sine wave is E_{YNR} leading E_{NR} by 30°. The phasor representing the e.m.f. acting from Y through N to R is the difference of the two phasors $E_{NR} - E_{NY}$.

$$E_{YNR} = E_{NR} - E_{NY}$$

Similarly in Fig. 4.7(d) $\quad E_{RNB} = E_{NB} - E_{NR}$

and in Fig. 4.7(e) $\quad E_{BNY} = E_{NY} - E_{NB}$

These phasors representing the e.m.f's which act between the terminals R and B, and B and Y are also included in the phasor diagram (Fig. 4.7(a)).

It will be observed that the magnitude of the induced e.m.f. acting between the two terminals Y and R is given by

$$E_{YNR} = 2 E_{NR} \cos 30°$$
$$= 2 E_{NR} \sqrt{3}/2 = 1\cdot732 E_{NR}$$

You will now understand why in the test carried out on the three-phase star-connected alternator the line voltage was 173 volts when the phase voltage was 100 volts.

The e.m.f. (E_L) induced between any two finishes of the coils is termed the line e.m.f. while the e.m.f. (E_{ph}) induced between the start and finish of a coil is termed the phase e.m.f.

$$E_L = \sqrt{3} E_{ph}$$

If the voltage of a three-phase supply is referred to as 400 volts then we understand that the r.m.s. voltage between the lines is 400 volts. The voltage between any line and neutral will be $400/\sqrt{3}$ = 230 volts.

Voltage drops and currents in a star-connected three-phase load

Experiment 4(ii) *To connect load to a three-phase star-connected alternator and to confirm by constructing phasor diagrams the voltages and currents in the system*

The alternator (Fig. 4.8) was loaded with three equal resistors connected in star, and the field excitation adjusted until the voltage across the red phase of the load was 100 volts. The load currents and voltages were then measured and the values tabulated.

It was observed that when the switch S_n in the neutral connector was opened there was no change in the readings of the instruments. It is clear from the results that with three similar resistors connected in

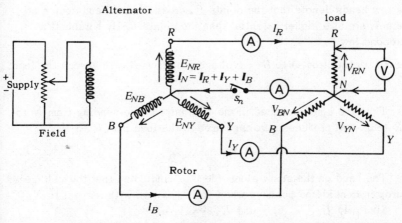

Fig. 4.8. Three-phase alternator with a balanced load.

V_{RN}	V_{YN}	V_{BN}	V_{RNY}	V_{YNB}	V_{BNR}	I_R	I_Y	I_B	I_N
(V)	(V)	(V)	(V)	(V)	(V)	(A)	(A)	(A)	(A)
100	100	100	173	173	173	0·9	0·9	0·9	0

star to a three-phase alternator the line currents are equal in magnitude and the voltage drops between the line terminals of the load are $\sqrt{3}$ times the phase voltage drops across the load.

Before constructing the phasor diagram you may find it helpful to consider a d.c. circuit with a battery supplying current to a resistor. (Fig. 4.9).

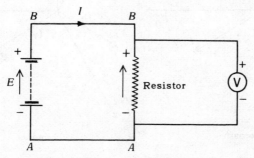

Fig. 4.9. E.M.F. and voltage drop in a d.c. circuit.

The battery is a source of e.m.f. (as are the phase windings of an alternator) and the e.m.f. acts within the battery from A to B. The e.m.f. produces a current I which flows through the load resistor with a consequent resulting voltage drop from B to A across the load. The arrow heads denote that the points B across the load and source of e.m.f. are at a higher potential than the points A. By Kirchhoff's second law,

e.m.f. acting from A to B = voltage drop from B to A across the load.
$$E_{AB} = V_{BA}$$

Thus the e.m.f. induced in the alternator phase winding from N to R (Fig. 4.8) produces a voltage drop in the load from R to N

$$E_{NR} = V_{RN}$$

The head on the arrow alongside V_{RN} indicates that the voltage drop is considered positive when R is positive with respect to N.

Similarly $E_{NY} = V_{YN}$ and $E_{NB} = V_{BN}$.

In Fig. 4.10(a) we have a phasor diagram representing the e.m.f's induced in the alternator windings and also between the alternator terminals.

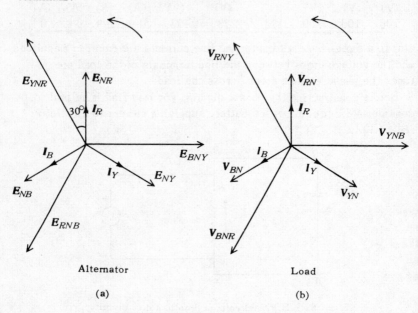

Fig. 4.10. Alternator and load phasor diagrams.

In Fig. 4.10(b) we have the phasor diagram for the corresponding voltage drops across the load. E.m.f's such as E_{YNR} produce voltage drops represented by V_{RNY}.

128

If the loading is purely resistive then, I_R, I_Y, and I_B will be in phase with V_{RN}, V_{YN} and V_{BN} respectively.

These current phasors are included in the diagrams and the phasor sum is zero. This accounts for the zero reading of the ammeter in the neutral connector. It is only when the load is balanced, that is when the impedances are identical in each phase of the load, that the current in the neutral will be zero.

Three-phase four-wire distribution system

This is a commonly used distribution system.

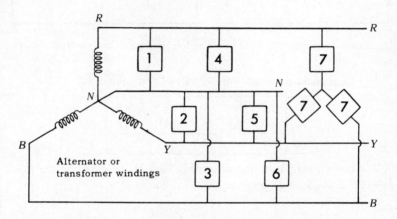

Fig. 4.11. Three-phase four-wire distribution system.

Consumers requiring single phase supplies are connected between one line and the neutral connector. The connections to six single phase consumers are shown in the figure. By connecting consumers to alternate R, Y and B lines an attempt is made to balance the loading on the three phases of the source. If a consumer requires a three phase supply then connections across the lines and neutral would be made as to consumer number 7.

Here is an experiment to illustrate this system of loading.

Experiment 4(iii) *To investigate the neutral-conductor current in a three-phase four-wire distribution system*

A six-pole alternator was loaded with resistors as in Fig. 4.12 and the field current adjusted until the alternator output was 200 volts when the machine was driven at 1000 rev/min.

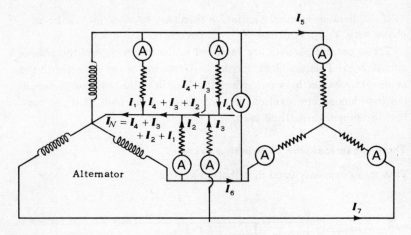

Fig. 4.12. Alternator supplying four single-phase loads and one three-phase load.

The instrument readings were observed and tabulated.

V	I_1	I_2	I_3	I_4	$I_5\, I_6\, I_7$	I_N
(V)	(A)	(A)	(A)	(A)	(A)	(A)
200	2·8	7·2	10·7	5·0	4·7	3·3

The three-phase load was balanced. This is indicated by the phase currents I_5, I_6 and I_7 being equal.

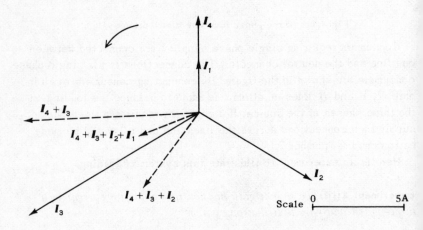

Fig. 4.13. Determining the neutral current.

Let us see if we can account for the current of 3·3 A returning to the alternator neutral point.

Since the loading is purely resistive the network currents will be in phase with appropriate voltages. I_1 and I_4 will be in phase with each other. Figure 4.13 shows the phasor diagram for the single phase loads.

The total neutral current will be given by,

$$I_N = I_1 + I_2 + I_3 + I_4$$

The graphical construction (Fig. 4.13) gives a value of about 3·4 A for I_N which agrees fairly closely with the reading of 3·3 A on the ammeter connected to the neutral of the alternator. Students should calculate the expected value of I_N by resolving the phasors I_1, I_2 and so on into components.

Connecting and disconnecting the balanced three-phase load does not make any difference to the resulting current in the neutral conductor. The potential of the load neutral point will be the same as the potential of the alternator neutral point.

Alternator with a rotating d.c. field system

We have assumed in dealing with three-phase alternating current theory that the conductors in which e.m.f's are induced in an alternator are on the rotating part of the machine. With almost all except the smallest machines it is usual to wind the coils in which the e.m.f's are induced,

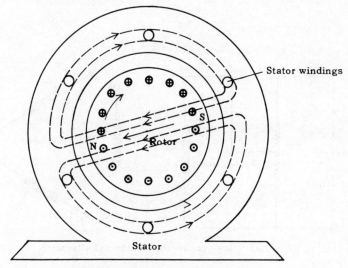

Fig. 4.14. Rotating field alternator.

on the stationary part of the machine which is referred to as the stator (Fig. 4.14). The magnetic field is provided by a system of coils on the rotor and this is driven by a prime mover.

With this arrangement certain advantages follow. When generating three-phase e.m.f's, if the conductors in which the e.m.f's are induced are on the rotor, a minimum of three slip rings will be required. If connections are to be made to the neutral point of the winding then a further slip ring will be necessary. When the d.c. field rotates however only two slip rings are required to lead the exciting current into and out of the rotor field windings.

For a given size of machine there is more space to accomodate conductors on the stator than on the rotor. Hence more conductors in which e.m.f's are to be induced may be provided if they are mounted on the stator. A higher e.m.f. can be generated and the machine output is increased. The additional space on the stator also enables us to provide the increased insulation necessary when higher voltages are generated.

Summation of the induced e.m.f's in a mesh-connected alternator

Mesh or delta connecting of a three-phase winding is an alternative to the star system of connections already dealt with. You will meet both these three-phase methods of connection when dealing with alternators, transformers and three-phase systems.

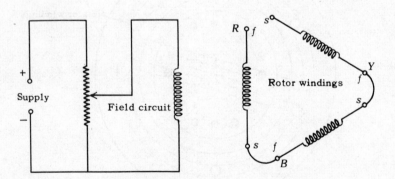

Fig. 4.15. Three-phase mesh-connected alternator.

Experiment 4(iv) *To investigate the e.m.f's induced in an alternator winding when the three phases are connected in mesh*

In this experiment the same three-phase alternator was used as in the previous test.

The alternator was driven at its rated speed and the field excitation adjusted until the e.m.f. induced in each phase winding was 100 volts. The red, yellow and blue phase windings were then connected in series and a voltmeter connected across the start of the yellow phase and finish of the red phase windings. The voltmeter indicated zero volts. We thus see that while e.m.f's are induced in the individual windings there is no resultant e.m.f. and hence no circulating current when the windings are connected in series.

Let us confirm this with a phasor diagram.

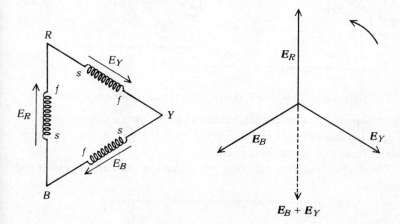

Fig. 4.16. Phasor representation of e.m.f's in a mesh-connected winding.

If you look carefully at this diagram you will see that we may combine the phasors E_B and E_Y which represent the e.m.f's in the red and yellow phases, to give a resultant phasor $E_B + E_Y$. This resultant e.m.f. is equal and opposite to E_R which represents the e.m.f. induced in the remaining phase. The phasors cancel and the resulting e.m.f. acting in the winding is zero.

Relationship between the line and phase currents of a mesh-connected alternator supplying a balanced three-phase load

Let us now consider a mesh-connected alternator supplying a load of pure resistance.

Experiment 4(v) *To connect a balanced load to a three-phase mesh-connected alternator and to confirm by constructing phasor diagrams the voltages and resulting currents in the system*

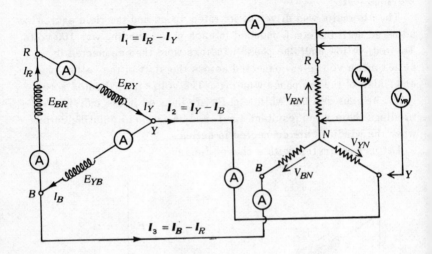

Fig. 4.17. Three-phase mesh-connected alternator with a balanced load.

The alternator was loaded with three equal resistors connected in star (Fig. 4.17) and the excitation adjusted until the voltage across the R and Y terminals of the load was 173 volts.

The various currents and voltages in the system were observed and tabulated.

V_{RN} V_{YN} V_{BN}	V_{RB} V_{YR} V_{BY}	I_R I_Y I_B	I_1 I_2 I_3
(V)	(V)	(A)	(A)
100	173	3·0	5·3

Figure 4.18(a) shows the phasor diagram for the alternator. E_{BR}, E_{RY} and E_{YB} represent the e.m.f's induced in the rotor coils which are displaced by $2\pi/3$ radians from each other. Since the load is purely resistive the corresponding voltages and currents must be in phase. Hence the current phasors I_R, I_Y and I_B may be drawn as shown.

Now the line currents between the alternator and the load are given by,

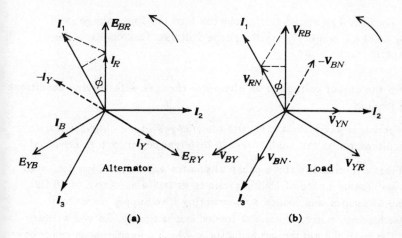

Fig. 4.18. Alternator and load phasor diagrams.

$$I_1 = I_R - I_Y$$
$$I_2 = I_Y - I_B$$
$$I_3 = I_B - I_R$$
$$I_1 = 2 \times I_R \cos 30° = \sqrt{3} I_R$$

Thus in a balanced mesh-connected alternator,

$$I_L = \sqrt{3} I_{ph}$$

This confirms theoretically the values for the line and phase currents as measured in the course of the experiment.

The voltage across any two lines in a mesh-connected system is the same as the phase voltage.

For the alternator

$$E_L = E_{ph}$$

In Fig. 4.18(b) we have the phasor diagram for the star-connected balanced and purely resistive load. The alternator induced e.m.f's E_{BR}, E_{RY} and E_{YB} are by Kirchhoff's second law equal to the load voltage drops V_{RB}, V_{YR} and V_{BY}. The line currents are I_1, I_2 and I_3. The load phase voltages V_{RN}, V_{YN} and V_{BN} will be in phase with these currents.

$$V_{RB} = V_{RN} - V_{BN}$$

hence
$$V_{RB} = 2 V_{RN} \cos 30° = \sqrt{3} V_{RN}$$

Thus for a balanced star-connected load the line voltage $V_L = \sqrt{3}\, V_{ph}$ where V_{ph} is the phase voltage. This confirms the experimental results.

How the output voltage of an alternator changes with varying conditions of loading

Experiment 4(vi) *To investigate the changes in the output voltage of an alternator as the load current at different power factors is varied*

In this experiment a three-phase alternator was driven by a d.c. motor at a constant speed of 1500 rev/min to generate an output of 50 Hz. The alternator was loaded by connecting it to supply power to a synchronous motor, which was loaded mechanically. As you will see in the next chapter the power factor at which a synchronous motor operates may be readily varied by adjusting its field excitation. Hence it is suitable for loading an alternator at varying power factors. An alternative method of loading would have been to use three-phase resistor, capacitor and inductor units.

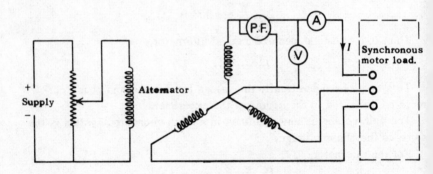

Fig. 4.19. Alternator loaded to a synchronous motor.

The alternator field regulator was adjusted until the output was 120 volts across each phase on no load. The regulator was left in this position throughout the tests.

The load on the synchronous motor was adjusted in steps to produce an increasing load on the alternator. By adjusting the synchronous motor field regulator the operating power factor was maintained constant at (a) 1·0, (b) 0·8 lagging and (c) 0·8 leading.

The instrument indications were as tabulated.

I (A)	P.F. ($\cos\phi$)	V_{ph} (V)
–	1·0	120
1·0	1·0	120
5·0	1·0	117
8·0	1·0	112
9·5	1·0	110
10·4	1·0	110
12·0	1·0	104
	lagging	
–	0·8	120
2·9	0·8	112
6·7	0·8	100
8·7	0·8	96
	leading	
–	0·8	120
3·0	0·8	128
4·0	0·8	130
6·6	0·8	132
8·6	0·8	134
10·0	0·8	136
13·1	0·8	140

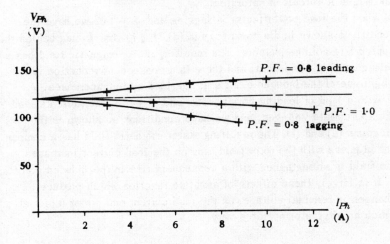

Fig. 4.20. Alternator output voltage load characteristics.

These observations were plotted as in Fig. 4.20 and this graph shows that the alternator terminal voltage falls with increasing loads of unity and lagging power factors, but for loads of leading power factor the terminal voltage rises. How can we account for these voltage variations?

You will recall how when load on a d.c. generator increases there is a fall in the output voltage caused by the armature circuit resistance. In an alternator we have a similar fall in output voltage as the load current increases, because of the alternator winding impedance.

A further reason for a fall in the output of a d.c. generator with increasing load was the effect of armature reaction. We also have a similar effect in an alternator.

Let us consider armature reaction in an alternator. Assume that the d.c. magnetic field rotates and induces e.m.f's in three single-turn stator coils displaced by 120°. (Fig. 4.21(a)).

At the instant shown in Fig. 4.21(a) the e.m.f. induced in coil A will be at its maximum positive value. When the rotor field has rotated through 120° the e.m.f. induced in coil C will then be at its maximum positive value. Hence the e.m.f's induced in coils A, B and C may be represented by the phasor diagram (Fig. 4.21(b)).

If the load power factor is unity the currents in the coils will be in phase with the voltages and the alternator stator winding produces a cross magnetising field as in (a) which will distort the main field. Resultant weakening of the main field will only occur if any part of the magnetic circuit is saturated.

When the load power factor is lagging and $\phi = 60°$ we have the conditions shown in diagrams (c) and (d). I_A, I_B and I_C lag behind the appropriate voltage phasors. The resulting stator magnetic field has a demagnetising component, and there is a consequent reduction in the magnitude of the induced e.m.f's as the load current increases.

With a load of leading power factor and $\phi = 60°$ the currents lead the appropriate voltages, and we have conditions as shown in the diagrams (e) and (f). The resulting stator magnetic field has a component in phase with the rotor field, and as the load current increases the field is strengthened with a consequent rise in the induced e.m.f's.

It is largely these effects of armature reaction which produce changes in terminal voltage, as the load current and power factor at which an alternator operates vary.

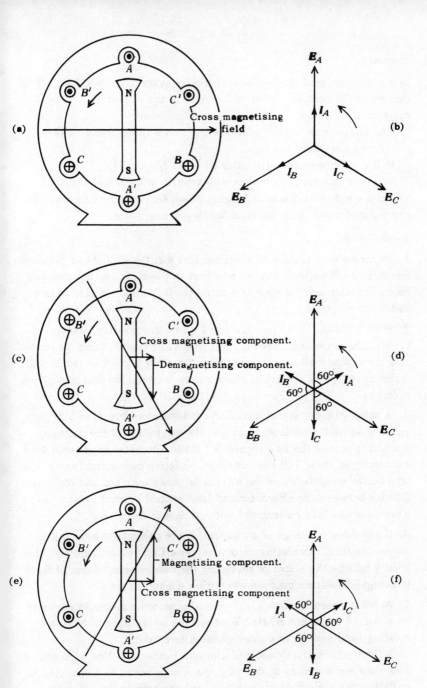

Fig. 4.21. Effect of the magnetic field produced by the load current on the d.c. field in an alternator.

Summary

In a star-connected three-phase, four-wire system $V_L = \sqrt{3}\, V_{ph}$. The current in the neutral conductor is equal to the phasor sum of the line currents.

The phasor sum of the induced e.m.f's in a three-phase mesh-connected alternator is zero.

In a mesh-connected alternator on load, $I_L = \sqrt{3}\, I_{ph}$.

Armature reaction in an alternator results in a decrease in the induced e.m.f. with loads of lagging power factor and an increase in the induced e.m.f. with loads of leading power factor.

Examples 4.1.

1. Describe with the aid of sketches how e.m.f's with phase displacements of $2\pi/3$ radians may be produced in three conductors mounted on an armature and rotating in a magnetic field. Show e.m.f. phasor and wave diagrams.

2. What is meant by a star-connected alternator winding? Sketch typical characteristics showing the phase and line voltages generated in a three-phase alternator winding as the field current is increased. Account for the relationship which exists between the line and phase voltages.

3. A three-phase star-connected alternator has an e.m.f. of 120 volts induced in each phase. What will be the magnitude of the voltages appearing across the line terminals? If the alternator supplies a load consisting of three 100-ohm resistors resistors connected in star what will be the magnitudes of the alternator phase currents and the current flowing between the alternator and load neutral points? Sketch the alternator and load current and voltage phasor diagrams.

4. If the three windings of an alternator are correctly connected in series the total circulating current is zero. Can you account for this? What would be the effect of reversing the connections to one of the windings? Illustrate your answer with a phasor diagram.

5. At what speed must an eight-pole alternator be driven to generate an e.m.f. of frequency 50 Hz? What are the advantages of having a rotating field system in a three-phase alternator?

A three-phase star-connected alternator generates line voltages of 200 volts and supplies a load of three 50-ohm resistors connected in delta (mesh). Calculate all voltages and currents in the alternator and load circuits.

6. Deduce the relationship in a three-phase balanced network between the line and phase voltages and between the line and phase currents, for both star and mesh systems of connection.

Each phase of a mesh-connected load consists of a coil having a resistance of 40 ohms and an inductance of 0·2 H. If the load is connected to a 415 volt, 50 Hz, three-phase, star-connected alternator determine the load phase currents, the line currents and the alternator phase voltage. Draw a phasor diagram to scale showing the relationship between the voltage and current in one branch of the load.

7. What is meant by the four-wire, three-phase distribution of electric power? How may consumers be provided with single and three-phase supplies from such a system?

Three consumers each take single-phase loads of 3 kW, 9 kW and 12 kW respectively from a 400 volt three-phase four-wire distribution system. Estimate the current in the supply neutral assuming that the loads are purely resistive.

8. Sketch a typical voltage/load current characteristic for a three-phase alternator supplying power to a load of leading power factor and account for the shape of the characteristic.

9. Describe how a load test at various power factors may be carried out on a three-phase alternator. What effect has the magnetic field produced by the load currents in the alternator windings, on the induced e.m.f's? Illustrate your answer with sketches.

Measuring power in a three-phase star-connected load using three wattmeters

Now that you understand how three-phase circuits may be connected in mesh or star we come to consider power in these systems.

Experiment 4(vii) *To use three wattmeters to measure the power input to a three-phase star-connected load*

The load consisted of a three-phase induction motor fitted with a mechanical brake which could be used to load the machine. Instruments were connected in the circuits as shown in Fig. 4.22.

The setting of the brake was adjusted until the motor line currents were 5 amperes, and the instrument readings were then observed and tabulated.

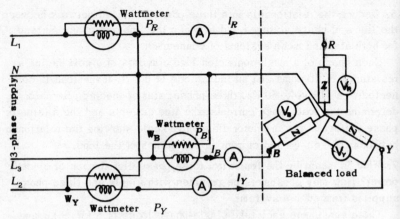

Fig. 4.22. Three wattmeter power measurement of a star-connected load.

| I_N | I_R | I_Y | I_B | V_R | V_Y | V_B | P_R | P_Y | P_B |
(A)	(A)	(A)	(A)	(V)	(V)	(V)	(W)	(W)	(W)
0	5	5	5	130	130	130	596	596	596

It is clear that the motor acts as a balanced load. The currents in the three lines are equal and consequently the neutral current is zero. The power is the same in each phase of the load.

$$\text{Total power} = V_R I_R \cos\phi_R + V_Y I_Y \cos\phi_Y + V_B I_B \cos\phi_B$$

$$= P_R + P_Y + P_B$$

$$= 596 + 596 + 596 \text{ watts}$$

$$= 1788 \text{ watts}$$

Power factor,
$$\cos\phi = \frac{P_R}{V_R \times I_R}$$

$$= \frac{596}{130 \times 5}$$

$$= 0.92$$

This three-wattmeter method of measuring power may be used with both balanced and unbalanced loads. If the loads are not equal in the three phases the wattmeters will indicate different values for the power in each phase.

If the load is known to be balanced only one wattmeter is necessary The total power (P) is then given by three times the reading of one wattmeter. If the wattmeter is connected in the red phase,

$$P = 3 V_R I_R \cos \phi_R$$
$$= 3 V_{ph} I_L \cos \phi$$
$$= 3 \frac{V_L}{\sqrt{3}} I_L \cos \phi$$
$$\text{Total power} = \sqrt{3} V_L I_L \cos \phi$$

This is an important expression for power in a balanced three-phase circuit.

Measuring power in a three-phase mesh-connected load using three-wattmeters

Experiment 4(viii) *To connect three wattmeters to measure the input power to a three-phase mesh-connected induction motor*

The instruments were connected as in Fig. 4.23 to a three-phase mesh-connected induction motor fitted with a brake for mechanical loading. The impedances Z represent the individual winding impedances.

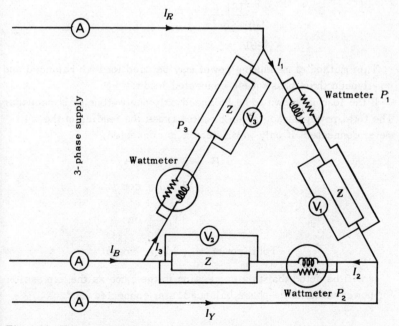

Fig. 4.23. Three-wattmeter power measurement of a mesh-connected load.

The load was adjusted until the current taken from the supply was 10 amperes in each line. All instrument readings were observed and tabulated.

I_1 (A)	I_2 (A)	I_3 (A)	I_R (A)	I_Y (A)	I_B (A)	V_1 (V)	V_2 (V)	V_3 (V)	P_1 (W)	P_2 (W)	P_3 (W)
5·75	5·75	5·75	10	10	10	226	226	226	1184	1184	1184

It is evident, since the three line currents are equal and the wattmeters give similar indications, that the motor provides a balanced load on the three-phase supply.

$$\begin{aligned}\text{Total power} &= V_1 I_1 \cos\phi_1 + V_2 I_2 \cos\phi_2 + V_3 I_3 \cos\phi_3 \\ &= P_1 + P_2 + P_3 \\ &= 1184 + 1184 + 1184 \\ &= 3552 \text{ watts.}\end{aligned}$$

Power factor,
$$\begin{aligned}\cos\phi &= \frac{P_1}{V_1 \times I_1} \\ &= \frac{1184}{226 \times 5\cdot75} \\ &= 0\cdot91\end{aligned}$$

This method of measuring power may be used for both balanced and unbalanced three-phase mesh-connected loads.

If the load is known to be balanced only one wattmeter is necessary. The total power is then given by three times the reading of the wattmeter connected. If only wattmeter P_1 is connected,

$$\begin{aligned}P &= 3 V_1 I_1 \cos\phi_1 \\ &= 3 V_L I_{ph} \cos\phi \\ &= 3 V_L \frac{I_L}{\sqrt{3}} \cos\phi \\ \text{Total power} &= \sqrt{3} V_L I_L \cos\phi\end{aligned}$$

It will be noted that this expression is the same as the expression for power in the three-phase balanced, star-connected load.

Connecting an artificial neutral point

If a load is known to be balanced and the neutral point of the supply is not available then the total power input may be determined by forming an artificial neutral point N and using only one wattmeter (Fig. 4.24).

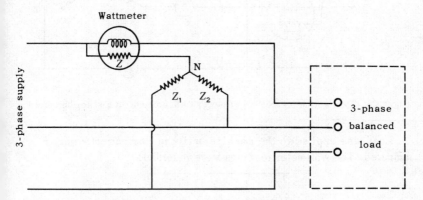

Fig. 4.24. Use of an artificial neutral point.

The external impedances Z_1 and Z_2 which are used to form the artificial neutral point must be identical with the wattmeter pressure circuit impedance Z. This is a method which may be used with both star and mesh-connected balanced loads.

The two-wattmeter method of measuring three-phase power in a star-connected load

The methods already dealt with to measure power in three-phase unbalanced loads require three wattmeters. Let us now have a look at a method of using two wattmeters to measure the power in three-phase loads.

Experiment 4(ix) *To measure the power input to a three-phase star-connected load using two wattmeters*

In this experiment the same machine was used as a load in **Experiment 4(vii)**, so that we could compare the results using different methods of power measurement. The wattmeters were connected as in Fig. 4.25. It will be observed that the neutral point of the load is not connected to the neutral point of the supply and hence at any instant $i_R + i_Y + i_B = 0$

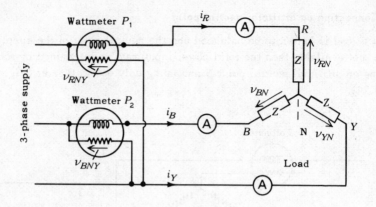

Fig. 4.25. Two wattmeter method of power measurement in a star-connected load.

Load was applied to the machine until the line currents were 5 amperes. The wattmeter readings were recorded.

I_R (A)	I_Y (A)	I_B (A)	P_1 (W)	P_2 (W)
5	5	5	652	1128

Now, with two wattmeters connected as in the circuit diagram it will be shown later that the total power is given by the sum of the readings.

$$\text{Total power} = P_1 + P_2$$
$$= 652 + 1128 \text{ watts}$$
$$= 1780 \text{ watts.}$$

This is in close agreement with the value of 1788 watts obtained using three wattmeters connected as in Fig. 4.22.

Let us have a look at the theory of the two wattmeter method of measuring three-phase power. It will be observed that the two wattmeter current elements carry the currents in two of the lines, while the wattmeter pressure elements are connected between the same line as the current element and the third supply line which does not include any current element.

Referring to Fig. 4.25

Instantaneous power in R phase = $i_R \nu_{RN}$

Instantaneous power in Y phase = $i_Y \nu_{YN}$

Instantaneous power in B phase = $i_B \nu_{BN}$

Total power at any instant = $i_R \nu_{RN} + i_Y \nu_{YN} + i_B \nu_{BN}$

Wattmeter P_1

Instantaneous current through the current element = i_R

Instantaneous p.d. across the voltage element = ν_{RNY}

$= \nu_{RN} - \nu_{YN}$

Instantaneous power measured by P_1 = $i_R(\nu_{RN} - \nu_{YN})$

Wattmeter P_2

Instantaneous current through the current element = i_B

Instantaneous p.d. across the voltage element = ν_{BNY}

$= \nu_{BN} - \nu_{YN}$

Instantaneous power measured by P_2 = $i_B(\nu_{BN} - \nu_{YN})$

Now consider the sum of two wattmeter readings.

$$P_1 + P_2 = i_R(\nu_{RN} - \nu_{YN}) + i_B(\nu_{BN} - \nu_{YN})$$

$$= i_R \nu_{RN} + i_B \nu_{BN} + \nu_{YN}(-i_R - i_B)$$

but $i_R + i_B + i_Y = 0$

$$\therefore i_Y = -i_R - i_B$$

$$\therefore P_1 + P_2 = i_R \nu_{RN} + i_B \nu_{BN} + i_Y \nu_{YN}$$

= total power in the three-phase load.

The steady deflection of a wattmeter indicates the mean value of the power. Hence the sum of the wattmeter readings gives the total mean power absorbed by the three-phase load. We have not assumed a balanced load or sine waves of current or voltage and therefore this two-wattmeter method may be used under all conditions. The wattmeters may be included in any two of the lines providing that one side of each pressure coil element is connected to the third line.

Effect of power factor on wattmeter indications when the load is balanced

In Fig. 4.26(a) we have the phasor diagram corresponding to the two wattmeter connection diagram shown in Fig. 4.25.

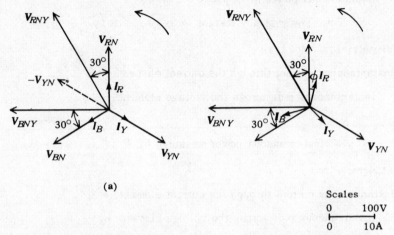

(a)

Scales
0   100V
0   10A

Fig. 4.26. Phasor diagrams corresponding to the two wattmeter power measurement circuit diagram in Fig. 4.25.

The phasor diagram is drawn in terms of phasors representing r.m.s. values of currents and voltages. It is drawn for load conditions of unity power factor. Since we are representing alternating quantities by phasors we are assuming sinusoidal waveforms.

Wattmeter P_1 will indicate $V_{RNY} I_R \cos 30°$ while wattmeter P_2 will indicate $V_{BNY} I_B \cos 30°$.

Writing V_{RNY} and V_{BNY} as V_L, and I_R and I_B as I_L we have,

$$P_1 = V_L I_L \cos 30°$$
and
$$P_2 = V_L I_L \cos 30°$$

Both wattmeters thus give similar indications under load conditions of unity power factor.

When the load is balanced, but the respective load currents lag the appropriate voltages by an angle ϕ, we have the conditions represented in the phasor diagram of Fig. 4.26(b).

The wattmeter indications will be related to,

$$P_1 = V_L I_L \cos(30° + \phi)$$
and
$$P_2 = V_L I_L \cos(30° - \phi)$$

When $\phi = 30°$

$$P_1 = V_L I_L \cos 60° = 0.5 V_L I_L$$
$$P_2 = V_L I_L \cos 0° = V_L I_L$$

The reading of one wattmeter is thus twice that of the other wattmeter.

When $\phi = 60°$

$$P_1 = V_L I_L \cos 90° = 0$$
$$P_2 = V_L I_L \cos(-30°) = 0.866 V_L I_L$$

One wattmeter therefore reads zero when the load power factor is $\cos 60°$.

When $\phi = 75°$

$$P_1 = V_L I_L \cos 105° = -0.259 V_L I_L$$
$$P_2 = V_L I_L \cos(-45°) = 0.707 V_L I_L$$

When ϕ exceeds 60° one wattmeter indicates a negative value. To enable this value to be determined it is usual to provide wattmeters with a switch in the circuit of the voltage element so that the pressure circuit connections may be reversed.

When $\phi = 90°$

$$P_1 = V_L I_L \cos 120° = -0.5 V_L I_L$$
$$P_2 = V_L I_L \cos(-60°) = 0.5 V_L I_L$$
$$P_1 + P_2 = 0$$

This is as we would expect since the power in a purely inductive three-phase load is zero.

Can we determine the power factor of a balanced load from the wattmeter readings?

Wattmeter indications will be given by,

$$P_1 = V_L I_L \cos(30° + \phi)$$
and
$$P_2 = V_L I_L \cos(30° - \phi)$$

Hence
$$P_1 + P_2 = V_L I_L (\cos 30° \cos \phi - \sin 30° \sin \phi + \cos 30° \cos \phi + \sin 30° \sin \phi)$$
$$= V_L I_L (2 \cos 30° \cos \phi) = V_L I_L \sqrt{3} \cos \phi \ldots \qquad (1)$$

and
$$P_1 - P_2 = V_L I_L (\cos 30° \cos \phi - \sin 30° \sin \phi - \cos 30° \cos \phi - \sin 30° \sin \phi)$$
$$= V_L I_L (-2 \sin 30° \sin \phi)$$
or
$$P_2 - P_1 = V_L I_L (2 \sin 30° \sin \phi) = V_L I_L \sin \phi \ldots \quad (2)$$

From equations (1) and (2),
$$\tan \phi = \frac{\sin \phi}{\cos \phi} = \sqrt{3}\left(\frac{P_2 - P_1}{P_1 + P_2}\right)$$

Using this expression to determine the power factor of the load in **Experiment 4(ix)** we have,
$$\tan \phi = \sqrt{3}\left(\frac{1128 - 652}{652 + 1128}\right) = 0.464$$
$$\therefore \phi = 24.9°$$
and $$\cos \phi = 0.91$$

This is very close agreement with the value of 0·92 obtained in **Experiment 4(vii)** for the same load.

The two wattmeter method of measuring power in a mesh-connected load

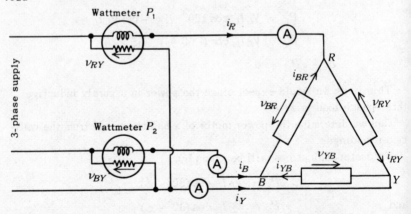

Fig. 4.27. Two wattmeter method of power measurement in a mesh-connected load.

Experiment 4(x) *To measure the power input to a three-phase mesh-connected load using two wattmeters*

In this experiment the same three-phase induction motor load was used as in **Experiment 4(viii)** when the total power was determined using three wattmeters.

Load was applied to the motor by means of the brake mechanism mounted on the shaft, until the line current was 10 amperes. The instrument readings were observed and tabulated.

I_R	I_Y	I_B	P_1	P_2
(A)	(A)	(A)	(W)	(W)
10	10	10	1270	2280

Now it will be seen later that the total power may be obtained from the sum of the two wattmeter readings, when the wattmeters are connected as shown.

$$\text{Total power} = P_1 + P_2$$
$$= 1270 + 2280 \text{ watts}$$
$$= 3550 \text{ watts.}$$

This is in close agreement with the value of 3552 watts obtained previously for the total power measured by three wattmeters with the same load.

Let us have a look at the theory of the two wattmeter method of power measurement in a mesh-connected three-phase load.

Referring to Fig. 4.27,

Wattmeter P_1

Instantaneous current through the current element $= i_{RY} - i_{BR}$

Instantaneous p.d. across the voltage element $= v_{RY}$

Instantaneous power measured by $P_1 = v_{RY}(i_{RY} - i_{BR})$

Wattmeter P_2

Instantaneous current through the current element $= i_{BR} - i_{YB}$

Instantaneous p.d. across the voltage element $= v_{BY}$

$$= -v_{YB}$$

Instantaneous power measured by $P = -v_{YB}(i_{BR} - i_{YB})$

Now consider the sum of the two wattmeter readings

$$P_1 + P_2 = v_{RY}i_{RY} + v_{YB}i_{YB} - v_{RY}i_{BR} - v_{YB}i_{BR}$$
$$= v_{RY}i_{RY} + v_{YB}i_{YB} + i_{BR}(-v_{RY} - v_{YB})$$

but
$$v_{RY} + v_{YB} + v_{BR} = 0$$
$$\therefore v_{BR} = -v_{RY} - v_{YB}$$
$$P_1 + P_2 = v_{RY}i_{RY} + v_{YB}i_{YB} + v_{BR}i_{BR}$$
$$= \text{total power in the three-phase load.}$$

As with the two wattmeter method applied to the star-connected load this does not assume balanced loading or sine waves and may therefore be used under all conditions.

The wattmeters may be included in any two lines providing that one side of each pressure coil element is connected to the third line.

Fig. 4.28(a) shows a phasor diagram corresponding to the two wattmeter connections in Fig. 4.27 when the load power factor is unity.

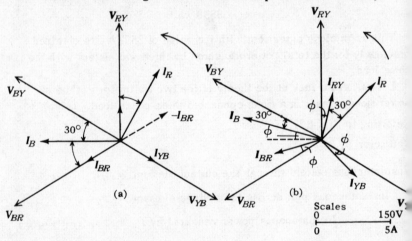

Fig. 4.28. Phasor diagrams corresponding to the two wattmeter power measurement circuit diagram in Fig. 4.27.

The diagram is drawn in terms of r.m.s. values of currents and voltages for a load of unity power factor. The use of phasors implies that sinusoidal waveforms are being considered.

The wattmeter indication will be given by,

$$P_1 = V_{RY}I_R \cos 30°$$

and
$$P_2 = V_{BY}I_B \cos 30°$$

Hence under conditions of balanced load and unity power factor both wattmeters will give similar indications.

Fig. 4.28(b) represents the conditions when the balanced load has a lagging power factor $\cos \phi$. By replacing V_{RY} and V_{BY} by V_L, and I_R and I_B by I_L it is seen that the indications of P_1 and P_2 will be given by,

$$P_1 = V_L I_L \cos(30 + \phi)$$

and

$$P_2 = V_L I_L \cos(30 - \phi)$$

These are similar to the expressions derived when dealing with the two wattmeter method of power measurement in balanced star-connected load circuits. The load power factor may consequently be determined from the expression

$$\tan \phi = \sqrt{3}\left(\frac{P_2 - P_1}{P_1 + P_2}\right)$$

Applying this expression to determine the power factor of the load in **Experiment 4(x)** we have,

$$\tan \phi = \sqrt{3}\left(\frac{2280 - 1270}{1270 + 2280}\right) = \sqrt{3}\frac{1010}{3550} = 0.494$$

Hence $\phi = 26.3°$ and $\cos \phi = 0.9$

This is in very close agreement with the value of 0.91 obtained in **Experiment 4(viii)** for the same load.

Summary

In a balanced three-phase star or mesh-connected load the power is given by $\sqrt{3}\, V_L I_L \cos \phi$.

Power in three-phase balanced and unbalanced three-wire circuits may be measured using two wattmeters. The total power is given by the sum of the wattmeter readings. When the load is balanced the power factor may be calculated from the wattmeter readings,

$$\tan \phi = \sqrt{3}\left(\frac{P_2 - P_1}{P_1 - P_2}\right)$$

When ϕ exceeds 60° one of the wattmeters will give a negative reading.

Examples 4.2.

1. How may the total power in a three-phase load be determined using three wattmeters? Sketch circuit diagrams showing the wattmeter connections when the load is (a) star and (b) mesh-connected.

What would be the effect on the wattmeter readings in each circuit if the load was balanced?

2. Show that for a balanced three-phase load of either star or mesh-connections the total power is given by $\sqrt{3}\, V_L I_L \cos\phi$.

3. How may an artificial neutral point be formed to enable the power input to a three-phase induction motor to be measured, when the neutral of the supply is not accessible? The load may be assumed to be balanced.

4. How may the power input to a three-phase star-connected load be measured using two wattmeters? Give a circuit diagram and explain how this method is applicable to balanced and unbalanced loads.

5. Show with the aid of diagrams how three-phase e.m.f's are generated in a rotating-field alternator.

A balanced three-phase load of three impedances connected in star is connected across a 400 volt three-phase 50 Hz alternator. Each impedance consists of a 40-ohm resistor in series with a 0·1 henry inductance. Calculate, (a) the alternator power output, (b) the line current and (c) the load phase voltage. Give a circuit diagram and show how one instrument could be connected to measure the alternator power output.

6. Three similar coils each having a resistance of 20 ohms and a reactance of 20 ohms are connected (a) in star and (b) in mesh to a 440 volt three-phase system. Calculate for each method of connection the kVA and power.

Determine also the current when one of the coils is disconnected.

7. A three-phase balanced load consists of three coils connected in star across a 200 volt three-phase 50 Hz supply. Each coil has a resistance of 30 ohms and an inductive reactance of 40 ohms. Two wattmeters are used to obtain the total power. One wattmeter has its current coil in the red line and the other in the blue line, the phase sequence being red, yellow, blue.

Construct a phasor diagram to scale showing the currents through the wattmeter current coils and also the potentials across the pressure circuit elements. Referring to this diagram calculate the readings of the wattmeters.

Delta to star and star to delta transformations

Occasions arise — particularly in the solving of three-phase network problems — where it is desirable that we should be able to replace three impedances connected in mesh or delta by three impedances connected in star. Transformations in the reverse direction from star to delta may also be required.

The two networks (Fig. 4.29) must be equal in that if we look into any two corresponding pairs of terminals then similar impedances will be presented by the network.

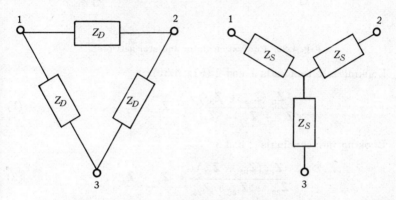

Fig. 4.29. Equivalent delta and star impedance networks.

Let us consider these circuits and the relationships between the impedances which will ensure that the desired conditions hold.

(a) Three equal impedances.

Looking into terminals 1 and 2 (Fig. 4.29).

$$\frac{Z_D(Z_D + Z_D)}{Z_D + (Z_D + Z_D)} = Z_S + Z_S$$

$$\frac{2 Z_D^2}{3 Z_D} = 2 Z_S$$

$$Z_S = \frac{Z_D}{3}$$

and

$$Z_D = 3 Z_S$$

(b) General case where the three impedances are not necessarily equal.

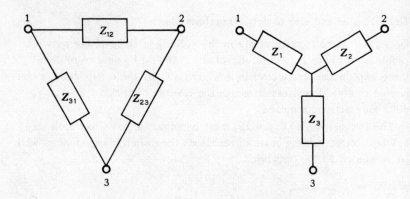

Fig. 4.30. Equivalent delta and star networks.

Looking into terminals 1 and 2 (Fig. 4.30).

$$\frac{Z_{12}(Z_{23} + Z_{31})}{Z_{12} + Z_{23} + Z_{31}} = Z_1 + Z_2 \ldots \quad (1)$$

Looking into terminals 2 and 3

$$\frac{Z_{23}(Z_{12} + Z_{31})}{Z_{23} + Z_{12} + Z_{23}} = Z_2 + Z_3 \ldots \quad (2)$$

Looking into terminals 3 and 1

$$\frac{Z_{31}(Z_{12} + Z_{23})}{Z_{31} + Z_{12} + Z_{23}} = Z_3 + Z_1 \ldots \quad (3)$$

Now, add equations (1) and (3) and subtract equation (2).

$$2Z_1 = \frac{2Z_{12}Z_{31}}{Z_{12} + Z_{23} + Z_{31}}$$

$$Z_1 = \frac{Z_{12}Z_{31}}{Z_{12} + Z_{23} + Z_{31}} \ldots \quad (4)$$

Similarly,

$$Z_2 = \frac{Z_{23}Z_{12}}{Z_{23} + Z_{31} + Z_{12}} \ldots \quad (5)$$

$$Z_3 = \frac{Z_{31}Z_{23}}{Z_{31} + Z_{12} + Z_{23}} \ldots \quad (6)$$

These expressions constitute the delta star transformation which may be expressed in words as follows: the impedance connected to one terminal of the star network is equal to the product of the two delta impedances connected to the same terminals in the delta network, divided by the sum of the three impedances in the delta network.

Let us now see how a transformation from star to delta may be brought about.

Consider equations (4), (5) and (6), multiplying equation (4) × (5),

$$Z_1 Z_2 = \frac{Z_{12}^2 Z_{31} Z_{23}}{(Z_{12} + Z_{23} + Z_{31})^2}$$

multiplying equation (5) × (6),

$$Z_2 Z_3 = \frac{Z_{23}^2 Z_{12} Z_{31}}{(Z_{12} + Z_{23} + Z_{31})^2}$$

multiplying equation (6) × (4),

$$Z_1 Z_3 = \frac{Z_{31}^2 Z_{23} Z_{12}}{(Z_{12} + Z_{23} + Z_{31})^2}$$

Adding,

$$Z_1 Z_2 + Z_2 Z_3 + Z_1 Z_3 = \frac{Z_{12} Z_{23} Z_{31} (Z_{12} + Z_{23} + Z_{31})}{(Z_{12} + Z_{23} + Z_{31})^2}$$

$$= \frac{Z_{12} Z_{23} Z_{31}}{Z_{12} + Z_{23} + Z_{31}}$$

From equation (6)

$$Z_3 = \frac{Z_{31} Z_{23}}{Z_{12} + Z_{23} + Z_{31}}$$

$$\therefore Z_1 Z_2 + Z_2 Z_3 + Z_1 Z_3 = Z_{12} Z_3$$

and

$$Z_{12} = \frac{Z_1 Z_2 + Z_2 Z_3 + Z_1 Z_3}{Z_3}$$

$$= Z_1 + Z_2 + \frac{Z_1 Z_2}{Z_3}$$

Similarly,

$$Z_{23} = \frac{Z_1 Z_2 + Z_2 Z_3 + Z_1 Z_3}{Z_1}$$

$$= Z_2 + Z_3 + \frac{Z_2 Z_3}{Z_1}$$

and

$$Z_{31} = \frac{Z_1 Z_2 + Z_2 Z_3 + Z_1 Z_3}{Z_2}$$

$$= Z_3 + Z_1 + \frac{Z_3 Z_1}{Z_2}$$

These equations enable us to convert from a star network to a delta network and may be expressed in words as follows: The single impedance in one path between two terminals of a delta network is equal to the sum of the impedances connected to the same terminals in the star network, together with an impedance given by the product of these two star impedances divided by the third star impedance.

Examples 4.3.

1. What do we mean by the equivalent star-circuit to a given mesh-network of resistors?

If three 100-ohm resistors are connected in mesh what would be the values of resistors in the equivalent star network? Derive any expression used.

2. Determine the branch impedances and power factors of a star network equivalent to a mesh network where each arm consists of a 30-ohm resistor in series with a 0·1 H inductor. The supply frequency is 50 Hz.

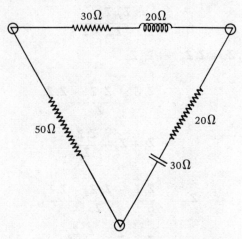

Fig. 4.31. Example 3.

3. Derive an expression for the star network equivalent to a mesh system when the branch impedances are not equal.

Calculate the values of components in the star network equivalent to the mesh network shown in Fig. 4.31.

4. Show how a given star network of three impedances may be replaced by a network of mesh connected impedances. If the star impedances each consist of a 30-ohm resistor in series with a 25-ohm inductive reactance determine the impedances in the equivalent mesh circuit.

5. A star impedance network has arms with impedances as shown in Fig. 4.32. Calculate the values of the impedances in the equivalent mesh network.

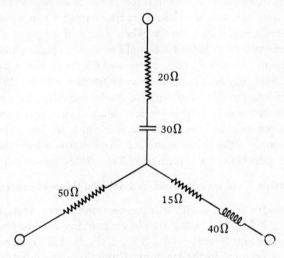

Fig. 4.32. Example 5.

5. Rotating Fields, Three-Phase and Single-Phase Alternating Current Motors, Power Factor Improvement and Electricity Tariffs

Introduction

You have now been introduced to the generation of three-phase e.m.f.'s and the measurement of power in three-phase systems. At this stage you might well ask why we should involve ourselves in what at first sight appears complicated theory, when we could 'get along' using single-phase systems. Indeed we could use only single-phase systems but we wish to 'get along' with maximum efficiency and it can be shown that alternators may be designed to operate more efficiently as three-phase machines than as single-phase machines. Also, by making three-phase systems available we are able to use a very simple motor which requires neither brush gear nor slip rings. We are going to consider this motor — the three-phase induction motor — but first we must study the production of a rotating field by a three-phase supply.

How a rotating field may be produced using a three-phase supply

Let us consider a stator with a cross-section as represented by the diagram Fig. 5.1(a). Six slots are shown carrying conductors which form three single turn coils AA', BB', CC'. At this stage we assume that there is no rotor in the machine. If the three stator coils are each connected between the neutral and different line terminals of a three-phase supply then currents will flow as represented by the phasors in (d), and the waves in (c) of Fig. 5.1.

At the instant X, the currents are as shown in Fig. 5.1(a). Coil A carries maximum current while the currents in coils B and C each have one half of this value. The six conductors of the three coils establish a magnetic field acting from right to left along a horizontal axis.

At the instant Y, the magnitudes and directions of the coil currents have changed (Fig. 5.1(b)). Current in coil B has now reached the positive maximum value and the currents in coils A and C are one half of this value. The direction of the field resulting from the conductors carrying current has changed and turned through an angle of $2\pi/3$ radians.

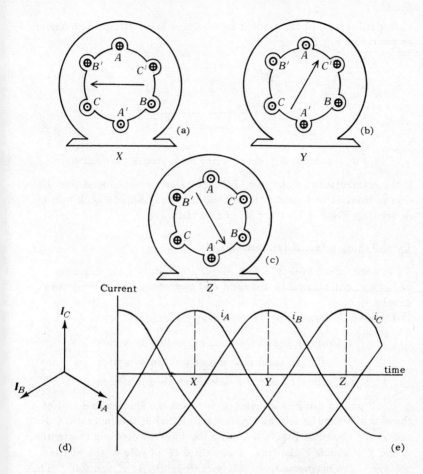

Fig. 5.1. Producing a three-phase rotating field.

At the instant Z, the current in coil C is at the positive maximum value and you will see that the direction of the resulting magnetic field is as shown in Fig. 5.1(c).

Now, we see from these diagrams that the magnetic field is in fact rotating. During one cycle of the supply voltage the field rotates through one complete revolution. By using this particular arrangement of coils connected to a three-phase supply we are able to produce a two pole rotating field.

Speed of rotating magnetic field (rev/s) = supply frequency (Hz)

This field-speed is termed the synchronous speed and depends entirely on the frequency of the supply.

A similar type of field could be obtained by rotating a bar magnet as shown in Fig. 5.2.

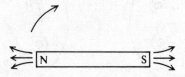

Fig. 5.2. Rotating field produced by a rotating bar magnet.

If the connections to any two of the coils are interchanged the direction of the field rotation is reversed. Students should check this by re-drawing Fig. 5.1 for the changed conditions.

Establishing a four-pole rotating field

To expand from a two-pole to a four-pole rotating field requires additional coils suitably arranged and connected to a three-phase supply.

Figure 5.3 shows how six coils may be arranged.

Coils AA' and DD' are connected in series,
Coils BB' and EE' are connected in series,
Coils CC' and FF' are connected in series.

Each pair of coils is connected between the line and neutral of a three-phase supply and the fields established at the instants X, Y and Z are shown in Fig. 5.3. During the interval between the instants X and Y — which is the time of one third of a cycle of the applied voltage — the magnetic field rotates through $\pi/3$ radians (60°). Thus during one complete cycle of the applied voltage the field rotates through π radians, or half a revolution.

The rotating field has four poles, but by suitable arrangement and connection of stator windings rotating fields with six, eight or more poles may be established.

The relationship between the rotational speed of the field, the number of pole pairs and the frequency of the supply is given by

$$n = \frac{f}{p}$$

where n is the rotational speed in rev/s, f is the frequency of the supply in Hz. p is the number of pole pairs.

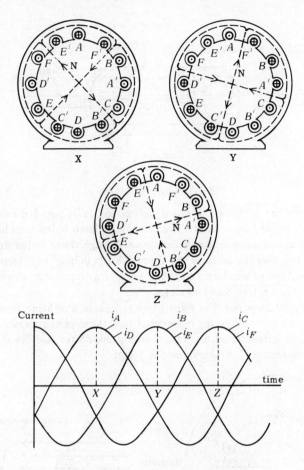

Fig. 5.3. Four-pole rotating field.

You will observe that this is the same expression as that which relates the frequency of the e.m.f.'s induced in an alternator winding, to the number of pole pairs in the field system and the speed at which the alternator is driven.

Now that you know how a rotating field may be produced when a three-phasor supply is connected to suitably arranged windings on a stator, you are ready to study the principle of operation of a three-phase induction motor.

How an induction motor operates

Let us consider a cage type of induction motor rotor as in Fig. 5.4.

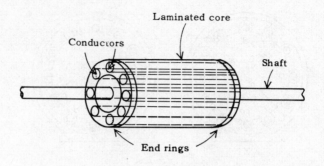

Fig. 5.4. Cage type rotor.

This rotor has a laminated construction and is assembled from steel stampings mounted on a shaft. The stampings have holes punched through them and copper bars which pass through these holes are welded to rings at the ends of the rotor. The 'winding' is referred to as a 'cage' type of winding and there are many closed circuits through the copper conductors and end rings.

We will now consider this rotor when it lies in a rotating field produced by a stator winding connected to a three-phase supply. Figure 5.5 is a simplified diagram showing the stator and the rotor. only one conductor P is shown on the rotor.

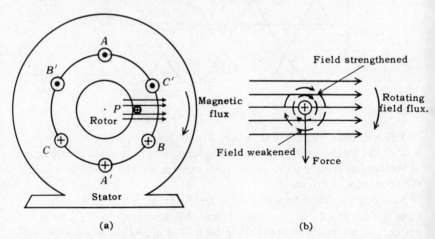

Fig. 5.5. Production of torque in an induction motor.

The stator windings carry three-phase currents and produce a rotating field which at the instant shown in Fig. 5.5(a) is sweeping downwards across the rotor conductor P. This magnetic field flux induces an e.m.f. in P and current flows in the rotor winding. The direction of the induced e.m.f. is shown in Fig. 5.5(b) and the rotating field is strengthened above P and weakened below P. The effect of this is that a force acts in a downward direction on the conductor P. Forces act on all rotor conductors as the rotating field sweeps past them, and the consequent torque produced on the rotor causes it to rotate.

The speed at which the rotor revolves can never quite reach the synchronous speed of the rotating field. If it did the rotating flux would no longer cut the conductors, and no e.m.f.'s would be induced to produce the circulating rotor currents which are essential before a torque can be developed. The difference between the synchronous speed and the rotor speed is termed the slip speed, and this is often expressed as a per unit or percentage of the synchronous speed.

$$n_s - n_R = s$$

$$\text{per unit slip} = \frac{n_s - n_R}{n_s}$$

$$\text{percentage slip} = \frac{n_s - n_R}{n_s} \times 100$$

$$n_s = \frac{f}{p}$$

n_s stator-field speed or synchronous speed (rev/s)

n_R rotor speed (rev/s)

s slip speed (rev/s)

f supply frequency (Hz)

p number of pole pairs for which the machine is wound.

You will appreciate that this three-phase cage-type induction motor is of very simple construction. It does not require a commutator, slip rings, or brush gear. Also it is not necessary to insulate the rotor conductors since the rotor currents will prefer to flow through the lower resistance paths of the cage winding rather than through the higher resistance paths of the iron laminations.

A total absence of any source of sparking makes this motor suitable for use in explosive atmospheres.

To reverse the direction of rotation it is only necessary to interchange the connections to two of the phases.

Starting an induction motor

When starting this type of motor the rotor is initially at rest and the rotating field produced by the stator windings cuts the rotor conductors at synchronous speed. The e.m.f.'s induced in the low resistance paths of the rotor produce large circulating currents. As you will realise later when dealing with transformers, large currents in the rotor windings of an induction motor — which may be compared with the secondary winding of a transformer — are accompanied by large circulating currents in the stator windings. To limit the rotor and stator currents to a safe value a reduced voltage is often applied to the stator circuits during starting. As the rotor speed increases the magnitude and frequenc of the induced e.m.f.'s decrease and the currents in the machine windings also decrease.

The reduced starting voltage may be obtained by supplying the machine through a transformer having adjustable tappings. As the motor speed rises the connections to the transformer are adjusted in steps by selector switches, until the full supply voltage is eventually applied across the motor stator windings.

An alternative method of starting is by means of star-delta connections. In this method the motor windings are connected in star during starting and then in delta (mesh) when the machine has reached its normal running speed. As you already know three impedances connected in star present a higher impedance than when connected in delta.

These methods of limiting the starting current are applied to machine with simple cage-type rotors. They have the disadvantage that by applying a reduced voltage during starting the starting torque developed by the motor is also reduced.

Instead of a cage-type rotor a wound rotor may be used. The rotor carries windings of insulated coils arranged in star with the free ends connected to slip rings. External resistance may be inserted into the rotor circuits during starting and this limits the rotor and hence the stator currents. As the rotor speed increases, the rotor resistance is gradually cut out until the rotor is finally short-circuited.

While a wound-rotor machine is more expensive than a cage-type it has the advantage that while starting the maximum torque may be obtained at low speeds.

Load test on a three-phase induction motor

Experiment 5(i) *To carry out a load test on a three-phase cage type induction motor*

In this experiment a three-phase mesh-connected cage-type induction motor was connected to a three-phase supply through a star-delta starter, as in Fig. 5.6.

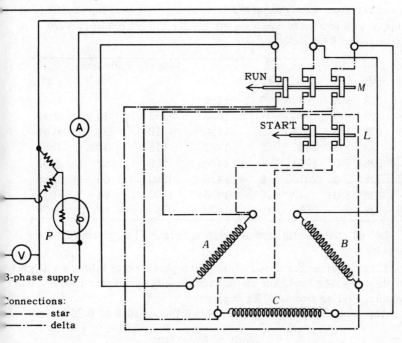

Fig. 5.6. Three phase induction motor load test.

The diagram shows only the main circuits of the starter. It does not include the overload and no-volt protective devices which are normally present in a motor starter. In the starting position the lower contactor L closes and the stator windings are connected in star. In the running position the contactor L opens and M closes to connect the windings in delta.

One disadvantage of the star-delta method of starting an induction motor is that during the changeover from the starting to the running positions of the controller the machine is disconnected from the supply, and the speed tends to fall momentarily. When the switch is closed into the running position the supply current should rise to three times the value immediately prior to the star connections being opened. Any fall in speed however during the changeover leads to a higher momentary increase in current, and this may lead to increased mechanical shock being transmitted to the unit driven by the motor.

The motor under test was a four-pole machine. It was loaded by a mechanical brake so that the torque developed under different conditions of loading could be determined. A single wattmeter and artificial neutral point resistors were connected to measure the power input. The load torque was increased in steps and the instrument readings were observed and tabulated.

V_L (V)	I_L (A)	Input (W)	T (Nm)	N rev/min.	Output (W)	Efficiency %	P.F. $\cos\phi$
228	5·8	780	0	1480	–	–	0·34
228	5·9	960	1	1480	155	16	0·41
228	6·0	1092	2	1480	310	28	0·46
228	6·4	1440	4	1470	616	43	0·57
228	7·0	1740	6	1440	908	52	0·63
228	7·6	2010	8	1430	1200	60	0·68
228	8·0	2340	10	1420	1490	64	0·74
228	8·7	2660	12	1420	1785	67	0·78

Let us consider the characteristics (Fig. 5.7) as plotted from these results.

It is seen that the speed of the three-phase motor falls only slightly with increasing load. The motor synchronous speed was 1500 rev/min when operating from a 50 Hz supply.

When delivering an output of 1·49 kW the speed was 1420 rev/min.

$$\text{Slip} = 80 \text{ rev/min}$$
$$\text{Percentage slip} = \frac{80}{1500} \times 100 = 5\cdot 3$$

The efficiency is seen to reach 67%. The motor losses include I^2R power losses in the stator and rotor windings. The stator core is subject to alternating magnetisation at the full supply frequency while the rotor core is subject to alternating magnetisation at slip frequency. Both the stator and rotor cores are therefore laminated to reduce eddy current power losses. Mechanical losses include friction at the bearings, and rotor windage loss.

The magnitude of the supply current changes from 5·8 A to 8·7 A over the range of loading considered. This change in current is accompanied by a change in operating power factor from 0·34 to almost 0·8. It is the resulting change in the active or power component of the supply current which provides the power necessary to meet increasing load demands on the machine.

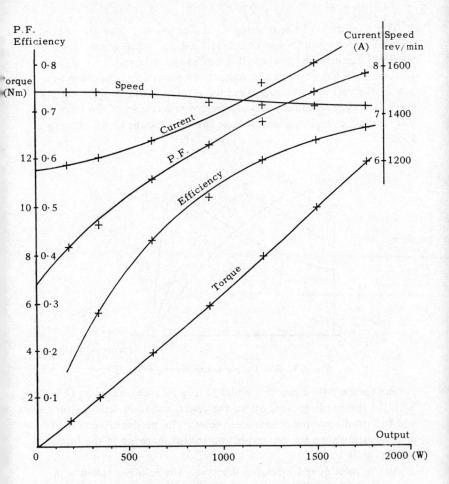

Fig. 5.7. Three-phase induction motor load characteristics.

Because of the almost constant speed characteristic the torque characteristic also increases almost uniformly with increasing load.

Torque in an induction motor

Let us consider how the torque developed varies as the rotor speed increases during starting. The torque (T) produced by a rotor depends on the power component of the rotor current ($I_R \cos \phi_R$), the magnetic flux density (B) of the stator field in which the conductors move, the number and length of the rotor conductors, and the rotor diameter.

For a particular machine

$$T \propto B \times I_R \times \cos \phi_R$$

The magnitude of the rotating field flux density depends on the applied voltage and this is constant. Cos ϕ_R is the power factor of the rotor circuits. A rotor winding has resistance which we will assume does not vary. At the instant of connecting a machine to a supply the rotor is stationary and the rotating field cuts the rotor at the full supply frequency. The inductive reactances of the rotor circuits are at their highest value. The rotor power factor cos ϕ_R is low.

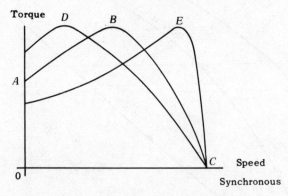

Fig. 5.8. How torque varies during starting.

The torque developed at standstill may be represented by OA (Fig. 5.8). As the motor gathers speed the reactance falls, cos ϕ_R increases and the total rotor impedance decreases. The product $I_R \cos \phi$ increases and hence we can expect an initial increase in the torque to a point such as B.

As the rotor speed increases however, the relative speed with which the rotating field cuts the rotor conductors falls and the reduced magnitude of the induced rotor e.m.f.'s causes I_R to decrease. Hence, even though the rotor reactance continues to fall and cos ϕ_R continues to rise there is a point such as B after which the product $I_R \times \cos \phi_R$ decreases. The torque characteristic then falls as shown from B to C.

The speed at which the maximum torque is developed during starting depends on the relative amounts of rotor resistance and rotor reactance. Increasing the rotor resistance produces a curve with maximum torque as shown at D, and reducing the rotor resistance gives a curve as shown by E.

We may control the speed at which the maximum torque is developed during starting by varying the amount of resistance in the rotor circuits. With a cage-type rotor the resistance is fixed and to enable us to vary rotor resistance a wound rotor must be used. Slip rings are provided through which additional rotor resistance may be connected during starting.

Here is an experiment on a wound-rotor slip-ring induction motor.

Experiment 5(ii) *To investigate the effects of rotor resistance on the torque characteristic of a three-phase induction motor*

The stator and rotor circuits of the machine were connected as in Fig. 5.9.

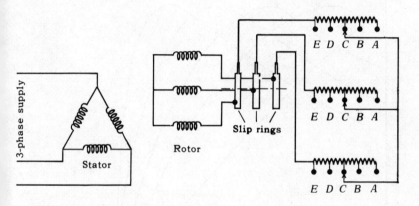

Fig. 5.9. Three-phase slip ring induction motor.

The stator windings were connected in delta and the rotor windings in star, with one end of each rotor phase connected to a slip ring. External resistors could be connected into the rotor circuits by a selector switch. This switch was interlocked with the main supply switch to ensure that the supply could only be connected to the stator when maximum additional resistance was included in the rotor circuits to limit the starting current.

The motor was loaded by a mechanical brake and the resistances in the rotor circuits varied in steps. At each step corresponding values of torque and speed were observed.

The results were as tabulated.

Selector position	T (Nm)	N rev/min.	T (Nm)	N rev/min.	T (Nm)	N rev/min.
A	4	1270	7	1140	14	820
B	4	1370	7	1310	14	1130
C	4	1415	7	1380	14	1270
D	4	1450	7	1526	14	1370
E	4	1470	7	1460	14	1430

(Resistance decreasing from A to E)

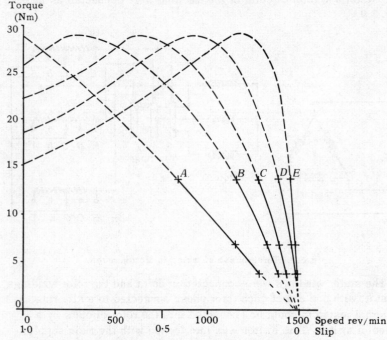

Fig. 5.10. Torque speed curves with varying rotor resistance.

The solid parts of the curves plotted in Fig. 5.10 represent these experimental results. It may be shown theoretically that the complete torque-speed curves have a general form as outlined by the broken lines. It is quite clear from the results that by including additional resistance in the rotor circuits higher torques are developed at lower speeds during starting.

Machines with double-cage rotors give an improved performance during starting, when compared with single cage machines. The outer cage winding near to the periphery of the rotor is a winding of low leakage reactance and high resistances.

An inner winding has comparatively higher leakage reactance and lower resistance. During starting the rotor currents are initially at the full supply frequency and are almost entirely confined to the outer winding of higher resistance and low reactance. This produces a high starting torque and also limits the starting current. At normal speeds the rotor currents flow largely in the inner cage of low resistance.

Varying the speed of an induction motor

The three-phase induction motor is essentially an almost constant speed machine and the speed may not be readily varied.

We have already seen how the supply starting currents of an induction motor can be limited by including variable resistance in the rotor circuits. In the previous experiment it has been seen that the rotor resistance has an effect on the speed at which the motor will develop a particular torque. Rotor resistance speed control of an induction motor is not however very efficient owing to the high I^2R power losses which are produced in the rotor circuits.

The coils on the stator of an induction motor may be so arranged that by suitable switching they produce rotating fields with different numbers of poles. This is known as pole changing and enables different speeds to be obtained from the same machine. Pole changing machines usually have cage type rotors, since the switching required to change the number of poles on both a wound rotor as well as the stator becomes complicated.

The Schrage a.c. commutator motor has been developed from the three-phase induction motor and speed control is smoothly variable over a wide range, but the theory of this machine is outside the scope of our present studies.

Summary

When a three-phase supply is applied to a suitable stator winding a rotating three-phase field may be produced.

The field speed, $n_s = f/p$ where f is the supply frequency and p the number of pole pairs for which the winding is designed.

If a cage type rotor is within this rotating field the e.m.f.'s induced produce the rotor circulating currents and a torque is developed. The rotor speed $n_r = n_s - s$ where n_s is the stator field or synchronous speed, and s is the slip speed.

To limit currents during starting a reduced voltage may be applied to the machine. Alternatively, a wound rotor with slip rings through which additional rotor resistance can be connected during starting may be employed.

As the load on a three-phase induction motor increases the speed falls only slightly. During starting the torque developed depends on the amount of resistance in the rotor circuits.

Examples 5.1.

1. Describe clearly with the aid of diagrams how a rotating field may be produced by a three-phase stator winding connected to a three-phase supply. How may the direction of the field rotation be reversed?

2. What is meant by synchronous speed? How could a four-pole rotating field be obtained from a stator winding connected to a three-phase supply? What would be the synchronous speed if the supply was of 60 Hz frequency?

3. Describe the construction of a cage-type induction motor. How is torque produced? What is meant by the slip speed? Explain with your reasons whether or not this slip can be zero.

A four-pole induction motor runs at 1460 rev/min on a 50 Hz supply. Calculate the percentage slip and the frequency of the e.m.f.'s induced in the rotor circuits.

4. Why is it necessary to provide an induction motor with a starter? Describe two methods of starting an induction motor.

5. What do you understand by the terms per-unit and percentage slip of an induction motor? A four-pole motor supplied at 415 volts, 3-phase 50 Hz has a rotor slip of 4%. Calculate the speed of the motor. What would be the speed of a six-pole alternator supplying power to the motor?

6. A 440 V, 3-phase star-connected induction motor has an output of 20 kW and a full load efficiency of 88%. If the operating power factor is 0·75 calculate the energy in kWh consumed during a period of four weeks if the motor operates on full load for four hours every day. What would be the supply line currents and the motor phase voltages?

How could the power input to this motor be determined if only one wattmeter was available? What assumption, if any, is made? Give a connection diagram.

7. Sketch the circuit diagram of a cage-type induction motor complete with a star-delta starter. Give typical characteristics showing torque, efficiency and speed to a base of output power. What are the losses which occur in this machine and how are they kept to a minimum? Are there any disadvantages in using star-delta starting?

8. A 440 volt 3-phase delta-connected induction motor runs at 1450 rev/min and drives a pump which raises 1000 litres of water per minute

to a height of 20 m. If the efficiency of the motor is 87% and that of the pump is 66% estimate the motor line and phase currents when the motor power factor is 0·8.

9. How does the torque-speed characteristic of a three-phase wound-rotor slip-ring induction motor vary during the starting of the motor?

How may the speed of an induction motor on load be varied by inserting a rotor resistance? Comment on any disadvantage of this method of speed control.

10. A 440 volt 3-phase mesh connected induction motor has an efficiency of 80% on full load and operates at a power factor of 0·75. Calculate the full load supply and motor winding currents if the output is 8 kW.

If this motor is supplied from a star-connected alternator calculate the alternator phase and line currents and voltages. How may the power input to the motor be measured when the supply neutral is not available?

11. What advantage has a double-cage rotor when compared with a single-cage rotor?

A three-phase cage type induction motor is connected in mesh and takes a line current of 20 amperes from a 415 volt three-phase supply at 50 Hz. The total input power is 4200 watts. Assume an overall efficiency of 85% for the motor and calculate (a) the current in each phase of the stator winding, (b) the power output of the motor and (c) the operating power factor.

12. How may three-phase power be measured using two wattmeters? What would be the effect of unbalanced loads on the summation of the readings?

A three-phase 440 V star-connected alternator supplies a balanced three-phase delta-connected induction motor operating at 0·7 power factor and having a full load efficiency of 87%. If the motor output is 30 hp (1 hp = 746 W) determine the motor and alternator phase and line currents and voltages.

13. A motor generator set comprising a three-phase 50 Hz cage type induction motor driving a shunt connected d.c. generator is to be installed. From the following data estimate the current taken by the motor on full load. Generator: Terminal voltage 220 volts. Full load current 40 A. Full load efficiency 74%. Motor: Line voltage 400 V. Power factor 0·8. Efficiency 75%.

14. Draw a complete circuit diagram showing how one wattmeter may be connected to measure the input power to a three-phase wound rotor induction motor when the neutral of the supply is available. Include

rotor-resistance and stator-control connections in your diagram. What assumption is made in using only one wattmeter?

If the motor is a six-pole machine having rotor currents of frequency 2·5 Hz and it is supplied from a four-pole alternator driven at 1500 rev/min calculate the rotor speed.

The synchronous motor

You have seen how a rotating field is produced when the winding on the stator of an induction motor is connected to a three-phase supply.

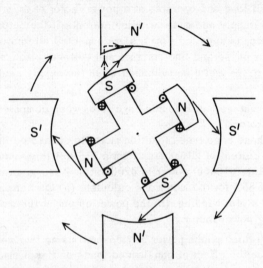

Fig. 5.11. How torque is produced in a synchronous motor.

In Fig. 5.11 the four magnetic poles N' and S' are the poles of a rotating field produced by a stator winding. Let us consider within this field a rotor with windings carrying direct current and producing four magnetic poles as N and S. If, by some external means, the rotor is brought up to the synchronous speed of the rotating field, the moving stator poles will attract the rotor poles and rotation will then continue at synchronous speed without further external assistance. The driving torque is produced by the rotor poles lying relatively slightly behind the stator poles. A component of the inclined force of attraction between neighbouring poles acts in the direction of rotation. This is the principle upon which the operation of a synchronous motor depends. Figure 5.11 shows a rotor with salient poles, but cylindrical rotors with suitable windings are also commonly used. Figure 5.12 represents a machine with a two-pole cylindrical rotor.

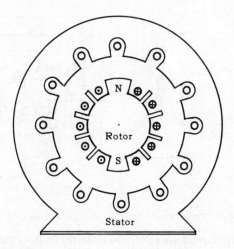

Fig. 5.12. Two-pole cylindrical rotor machine.

The rotor speed may be raised to the synchronous speed by using either an auxillary motor or by arranging for the machine to start as an induction motor. When the rotor has almost reached synchronous speed a d.c. supply is connected to the rotor windings which then establish magnetic poles, and the rotor jumps into a locked position with the rotating stator magnetic field. This type of machine which starts as an induction motor is termed a synchronous induction motor.

A cage type of rotor winding may alternatively be provided in addition to the d.c. winding. During starting, torque is established by the cage winding and steps must be taken to limit the supply current.

The following experiment is based on a wound rotor machine, and the action of the machine involves almost all topics you have already considered.

Experiment 5(iii) *To connect a three-phase synchronous induction motor and to observe the effect of varying the field excitation when the load is constant*

The machine was connected as in Fig. 5.13.

The stator windings were connected to a three-phase supply and arranged to produce a six-pole rotating field. During starting as an induction motor, the supply current was limited by including additional resistance in the rotor circuits. A d.c. shunt connected generator mounted on the rotor shaft was connected in the rotor circuit. As the rotor speed increased the rotor resistance was reduced. When the rotor had almost reached synchronous speed the d.c. injected by the

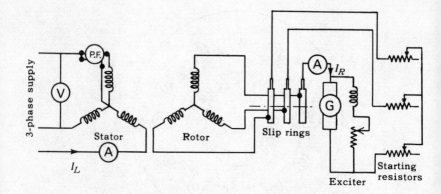

Fig. 5.13. Three-phase synchronous induction motor.

exciter was sufficient to give definite polarity to the rotor, and the rotor jumped into step with the stator field.

When running as a synchronous motor one phase of the rotor windings carries the full exciter current, while the other two phases each carry one half of the current in the reverse direction. This is equivalent to the conditions which exist in a stator or rotor winding connected to a three-phase supply, at the instant when the rotor currents flow as shown in Fig. 5.14. The rotor excitation current may be adjusted by varying the d.c. generator field regulating resistor.

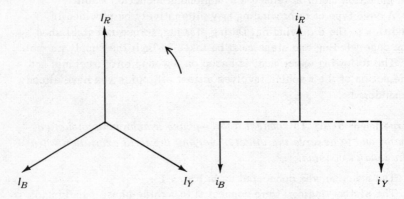

Fig. 5.14. Instantaneous values of rotor currents.

The motor was started on no load and when running at synchronous speed the rotor field current was adjusted until the supply line current was at a minimum value. Any further increase or decrease of

178

exciter current was observed to produce an increase in the supply current. Mechanical loading was then applied to the motor until it was developing one half full load torque. This load was maintained constant throughout the remainder of the test, and the effect of varying the rotor current was observed.

The instrument readings under both load and no load conditions were tabulated.

No load. V_L = 200 V

I_R (A)	2	3	4	5	6	7	8	9	10	11	12
I_L (A)	9·6	7·6	5·8	4·4	3·0	3·4	4·5	6·4	8·5	10·5	12·9
P.F. $\cos \phi$	0·2	0·3	0·38	0·5	0·75	0·66	0·46	0·34	0·26	0·2	0·16
	\multicolumn{5}{c}{P.F. lagging →}	← P.F. loading					

One half full load torque (T = 12·5 Nm) V_L = 200 V

I_R (A)	3	4	5	6	7	8	9	10	11	12
I_L (A)	11·0	9·5	7·0	6·4	6·5	7·4	8·7	10·2	12·4	14·3
P.F. $\cos \phi$	0·7	0·8	0·94	0·98	1·0	0·93	0·84	0·73	0·64	0·56
	P.F. lagging →				← P.F. loading					

Characteristics showing the variation in supply current and the operating power factor were plotted as in Fig. 5.15.

It will be observed that by suitably increasing the excitation a synchronous motor may be made to take a leading current from the supply.

The d.c. field rotates mechanically at synchronous speed and as it cuts the stator conductors it induced e.m.f.'s of supply frequency in these windings. When the motor is delivering a constant output we may assume that the power ($\sqrt{3}\ V_L I_L \cos \phi$) taken from the supply is constant. Since the applied voltage is fixed, any change in e.m.f.'s induced in the stator windings can only effect the product $I_L \times \cos \phi$.

Hence the power factor at which the motor operates depends on the d.c. field excitation.

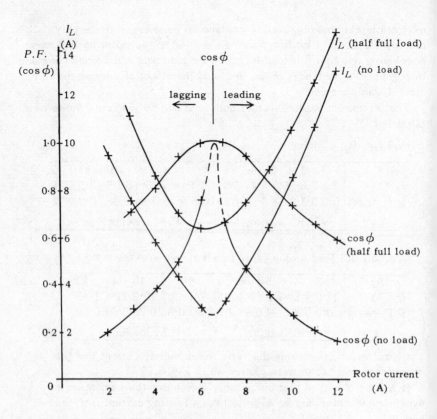

Fig. 5.15. Synchronous motor characteristics.

Producing a rotating field from a two-phase supply and the principle of the single-phase induction motor

You have already seen how a rotating field may be produced from a three-phase supply. Let us now consider how a rotating field may be obtained from a two-phase supply.

Figure 5.16(a) shows the current waves of a two phase supply where i_A leads i_B by $\pi/2$ radians. Let these currents flow in two single turn stator coils AA' and BB' which are also displaced by $\pi/2$ radians.

Considering the instants W, X, Y and Z it is seen that the combined magnetic field produced by the coils will be as represented in the diagrams (b) (c) (d) and (e) of Fig. 5.16. This is a rotating field making one complete revolution during one cycle of the alternating supply.

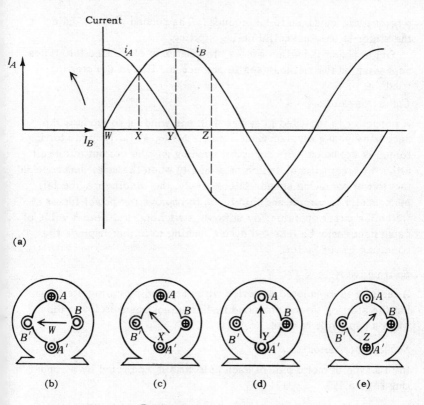

Fig. 5.16. Producing a two-phase rotating field.

Now, if a cage type of rotor is placed in the rotating field a torque will be produced on the rotor in the same manner as that produced by a three-phase rotating field.

How can we obtain a two-phase supply?

Our requirement is for two currents with a phase difference of $\pi/2$ radians. These may be obtained from a single-phase supply by connecting two coils in parallel. One partially inductive coil is connected directly across the supply and the other coil has additional resistance or reactance connected in series with it.

What happens if one phase (coil) is disconnected when the rotor has reached normal running speed?

The phase which remains connected will produce a magnetic field with a constant direction but of a pulsating nature. This field will induce e.m.f.'s and resulting currents in the cage rotor. There will be a phase displacement between the stator and rotor magnetic fluxes and

181

a torque will continue to be produced. The second phase winding on the stator is only essential during starting.

Single phase induction motors are commonly classified into types, depending on the method used to produce the flux in the starting winding.

Capacitor motors

A capacitor is included in series with one winding to produce the necessary phase displacement during starting. A fairly high starting torque is produced. The capacitor winding may be cut out of circuit either by a centrifugal switch or manually when the motor has reached the normal operating speed. Alternatively, the winding may be left permanently in circuit and this tends to improve the power factor at which the motor operates. By suitable switching, a different value of capacitance may be inserted during running to further improve the operating power factor.

Split-phase motors

A resistor is connected in series with one winding during starting but the starting torque produced by this arrangement is lower than when a capacitor is used.

Shaded-pole motor

In this type of motor part of each pole limb is encircled by a copper ring (Fig. 5.17).

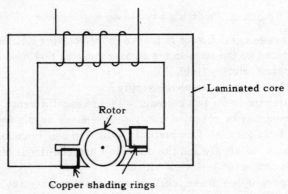

Fig. 5.17. Shaded-pole motor.

The alternating flux through the ring induces an e.m.f. and the resulting current causes the flux within the ring to be displaced from the main flux. This is sufficient to produce a motor which is self starting on light loads.

Single-phase a.c. series motor

This motor is similar in construction to the d.c. series motor but both the armature core and the yoke are laminated to reduce eddy current losses. During alternate half cycles of the supply current the field flux and armature currents both change directions to produce a unidirectional torque. On light loads the speed may become dangerosuly high because of the weak field. In the governor controlled type of motor a centrifugal switch is provided to insert additional resistance in series with the motor when the speed exceeds a certain value. Capacitors may be connected across the brushes to reduce sparking.

These motors are often termed universal motors and are suitable for operation from both a.c. and d.c. sources. As fractional-horse-power motors they are commonly used in vacuum cleaners. Large units of over 1000 kW rating operating on 16·77 Hz single-phase supplies have been used in traction work for which their high starting torque makes them particularly suitable.

Summary

A synchronous motor rotates at the synchronous speed of the rotating flux established by the stator windings. The rotor winding must be supplied with direct current, and the rotor must be brought up to synchronous speed during starting. The excitation may be obtained from a small d.c. generator mounted on the motor shaft. By varying the excitation current, the power factor at which the motor operates may be varied. The synchronous induction motor starts as a slip-ring induction motor and then runs as a synchronous motor.

A rotating field may be produced by phase splitting a single-phase supply and this is the basis upon which the single-phase induction motor operates. The a.c. series motor operates in a similar manner to the d.c. series motor, but both rotor and stator cores are laminated.

Examples 5.2.

1. Explain the principle of action of a three-phase synchronous motor and the essential difference between this motor and a three-phase cage-type induction motor.

2. Why are three-phase synchronous motors sometimes used in preference to induction motors? Sketch the circuit diagram for a synchronous induction motor.

3. Explain the statement that 'the synchronous motor may be made to operate with the characteristics of a capacitor'.

4. How may a rotating field be produced from a two-phase supply? What practical use is made of this principle?

5. Describe three types of single-phase motor which depend for their operation on rotating magnetic fields.

6. How is torque produced in an a,c. series motor? How do universal series motors differ from those intended solely for use in d.c. systems?

Power factor improvement

Alternating current load circuits with inductive reactance are much more common than those with capacitive reactance. All a.c. motors include coils with which alternating magnetic fluxes link and hence act as partially inductive loads.

Let us consider as an example a single-phase motor which has an input of 2000 watts on full load when operating from a 250 volt 50 Hz supply.

Assuming that the motor has an efficiency of 80% then the useful power output will be $2000 \times 80/100 = 1600$ watts.

If the motor behaved as a pure resistor the current taken from the supply would be $2000 W/250 V = 8$ amperes and the operating power factor would be unity. If however the motor has partially inductive windings then for the same output, input and efficiency, the current taken from the supply would be more than 8 amperes. With a power factor of 0·5 the current would be given by

$$I = \frac{P}{V \cos \phi} = \frac{2000}{250 \times 0·5} = 16 \text{ amperes}$$

Thus the motor operating at a lower power factor, but taking the same power and giving the same output as the motor operating at the higher power factor, requires a higher supply current.

Unless larger cables are used to supply the motor operating at the lower power factor, then the cable power losses ($I^2 R$) and also the cable impedance voltage drop ($I Z$) will increase. To provide heavier cables involves increased cost, and if a consumer has a large number of machines operating at comparatively low power factors then the electricity supply authority may have to meet considerably increased supply costs. It is desirable therefore that a consumer should operate with a power factor as near to unity as possible.

What can we do to improve the power factor when the load current is lagging behind the voltage? The basic answer is quite simple — the consumer installs apparatus which will take a leading current to cancel out the lagging component of the load current.

Here is an experiment to illustrate the effects in improving a low power factor.

Experiment 5(iv) *To improve the power factor of a load by using static capacitors*

In this experiment an artificial load which included resistance and inductive reactance was connected as in Fig. 5.18.

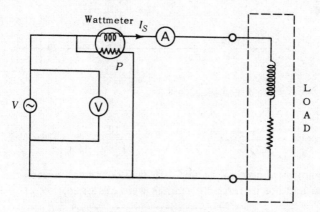

Fig. 5.18. Determining the power factor of a load.

The instrument readings were observed and the power factor calculated.

V (V)	I_s (A)	P (W)	$\cos \phi = P/VI_s$
200	5·6	625	0·558

In Fig. 5.19. I_L is the phasor for the load current. It lags the supply voltage by an angle such that $\cos \phi = 0.558$.

If a capacitor is connected across the supply so that it takes a current I_R equal in magnitude to the length AB on the phasor diagram, then the resultant supply current will be represented by OB in phase with V.

From the phasor diagram,

$$AB = I_L \sin \phi$$
$$= 5\cdot 6 \times 0\cdot 83 = 4\cdot 65 \, \text{A}$$

Hence,

$$X_c = \frac{V}{I_c} = \frac{200 \, \text{V}}{4\cdot 65 \, \text{A}} = 43 \, \Omega$$

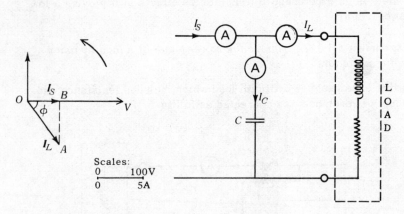

Fig. 5.19. Correcting power factor.

Now,
$$X_c = \frac{1}{2\pi f C} \therefore C = \frac{1}{2\pi f X_c} = \frac{10^6}{314 \times 43} = 74\mu F$$

A capacitor of nominal value $74\mu F$ was connected across the supply terminals and the instrument readings were observed.

V (V)	I_L (A)	I_s (A)	I_c (A)	P (W)
200	5·6	3·25	4·55	625

The new power factor presented to the supply is,
$$\cos \phi = \frac{\text{watts}}{\text{voltamperes}} = \frac{625}{200 \times 3 \cdot 25} = 0 \cdot 98$$

It will be observed that in correcting the power factor the supply current has been reduced from 5·6 A to 3·25 A.

Since the capacitor chosen to correct the power factor had only a nominal value of $74\mu F$ this probably accounts for the small discrepancy from unity of the final power factor actually achieved.

This experiment has illustrated how the power factor of a single load may be improved, and consequently the current demand on the supply reduced. If we take this a stage further and improve the power factor at which an entire factory is operating, then this means that for the same losses in the supply cables a greater load may be supplied. Alternatively, with an improved power factor and the same losses smaller cables are required to supply the same load, with a consequent saving in distribution costs.

The amount of capacitance required will vary depending on the load conditions and it may be desirable for automatic switching of capacitor units to take place when the load changes. Switching may be carried out using contactors operated through relays, which are sensitive to the reactive component of the load current.

Other devices used to improve power factor include synchronous motors which may be made to operate at leading power factors by suitable adjustment of their field currents, and also specialised machines which have been developed to supply reactive volt amperes.

Power factor correction is often applied to three-phase circuits and here is an example to illustrate this.

Example

A three-phase 440 V 50 Hz induction motor operates on full load at a power factor of 0·79 and the output is 40 kW. If the efficiency is 78% calculate the supply current. What values of capacitance connected in (a) star and (b) in mesh would improve the power factor to unity?

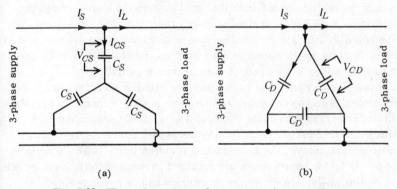

Fig. 5.20. Three-phase power factor correction capacitors.

Watts input $= 40\,000 \times \dfrac{100}{78}$

$$P = \sqrt{3}\, V_L\, I_L \cos\phi$$

$$\therefore I_L = \dfrac{P}{\sqrt{3}\, V_L \cos\phi} = \dfrac{40\,000}{\sqrt{3} \times 440 \times 0{\cdot}79} \times \dfrac{100}{78} = 85{\cdot}2\,\text{A}$$

Star connections Fig. 5.20 (a).

$$V_{cs} = \dfrac{440}{\sqrt{3}} = 254\,\text{V}$$

$$I_{cs} = I_L \sin\phi = 85{\cdot}2 \times 0{\cdot}615 = 52{\cdot}4\,\text{A}$$

$$X_{cs} = \frac{V}{I_{cs}} = \frac{254}{52\cdot 4} = 4\cdot 85\,\Omega$$

$$C_s = \frac{1}{2\pi f X_{cs}} = \frac{10^6}{314 \times 4\cdot 85} = 656\,\mu\text{F}$$

Delta connections (Fig. 5.20 (b)).

$$V_{cs} = 440\,\text{V}$$

$$I_{cs} = \frac{I_L \sin\phi}{\sqrt{3}} = \frac{52\cdot 4}{\sqrt{3}} = 30\cdot 2\,\text{A}$$

$$X_{cd} = \frac{V_{cd}}{I_{cd}} = \frac{440}{30\cdot 2} = 14\cdot 6\,\Omega$$

$$C_D = \frac{1}{2\pi f X_{CD}} = \frac{10^6}{314 \times 14\cdot 6} = 218\cdot 5\,\mu\text{F}$$

This example shows that when three-phase power factor correction capacitors are connected in delta, smaller currents pass through the capacitors with higher applied voltages than when the capacitors are connected in star. Hence smaller values of capacitance are required in delta. In some applications voltage step-up transformers may be used to further reduce the amount of capacitance required.

Power factor correction capacitors are usually oil impregnated and oil cooled. The power losses are very low and the capacitors operate with efficiencies exceeding 99·8% and only 15°C temperature rise. With no moving parts except for any necessary contactor selector switches and relays, the maintenance required is minimal. Capacitors for power factor improvement are provided in many installations including rolling mills, cement works, breweries and in motor vehicle assembly plants.

The output of an alternator or transformer is limited by the temperature rise which may be permitted. This in turn limits the current output from the machine. If a consumer is operating at a low power factor then he makes a higher demand on the available current carrying capacity of the system (cables, transformers, alternators and so on) than a consumer who takes the same power at a higher power factor Consumers are therefore encouraged to operate at a power factor as close to unity as possible by applying tariffs which penalise those who operate at low overall power factors.

Let us now consider the basis on which electricity tariffs are compiled.

Electricity tariffs

Here are two electricity tariffs.

(a) *Single-phase monthly service tariff*

Consumption		Charge
First	20 kWh or less	£1·00 (minimum charge)
Next	40 kWh	0·85 p per kWh.
Next	40 kWh	0·7 p per kWh.
Next	100 kWh	0·53 p per kWh.
All over 200 kWh		0·33 p per kWh.

Penalty charge for welding apparatus.

All welding apparatus will be charged for additionally at £1 per kVA input per month unless connected through a motor generator set. If the consumer has installed capacitors for power factor improvement the extra charge will be modified according to the degree of correction.

(b) *Three-phase monthly tariff with maximum demand metering*

Two-part tariff:

Part (a) Maximum demand charge.

First 75 kVA or less £75
All additional kVA of maximum demand at £1 per kVA.
The maximum demand meter will be read monthly.

Part (b) Energy charge.

Consumption		Charge
First	5000 kWh	0·70 p per kWh
Next	5000 kWh	0·63 p per kWh
Next	40 000 kWh	0·52 p per kWh
Next	100 000 kWh	0·42 p per kWh
Next	850 000 kWh	0·33 p per kWh
All over	1 000 000 kWh	0·29 p per kWh.

Discount.

Consumers supplied at high voltage will receive the following discounts:

Supply	Discount
3300 V, 6600 V, 11 000 V	2%
33 000 V, 66 000 V	3%
132 000 V	5%

The arrangement of these two tariffs is typical of the tariffs used by electricity undertakings. Although individual charges may vary considerably there are similar underlying principles upon which the tariffs are based.

In providing a supply of electricity an authority must meet certain costs and it is essential that the revenue collected from the sale of the electrical energy covers these costs and provides the requisite margin of profit.

The costs of supply may be divided into two broad groups.

(a) *Fixed charges*

Capital must be borrowed to purchase land, plant and buildings, and an annual allowance must be made to cover depreciation. Hence we have:

(i) interest and depreciation charges on plant including generators, transformers, switchgear, transmission lines, generating and substation buildings

(ii) costs of the constant losses which occur in the plant.

(iii) a portion of the administrative charges.

(b) *Running charges*

These depend largely on the amount of electrical energy supplied and the cost of fuel is the main item in other than hydroelectric stations. Other charges which come into this section include the cost of lubricating oils, cooling water, maintenance and labour charges, and also variable energy losses which occur in the plant.

The electricity undertaking has to meet the fixed costs even though it may be generating very little power. It is to cover the fixed costs that the tariffs make a high initial charge to consumers.

The single phase domestic consumer pays at a higher rate for the initial block of kWh consumed. The industrial consumer is charged heavily for the maximum kVA demand made on the supply undertaking. The capital costs incurred in providing this supply are related to the maximum demand. By basing the fixed charges on kVA rather than on kW the consumer is encouraged to operate at a high power factor, and this reduces his maximum current demand on the system.

Consumers who are prepared to take supplies at periods when there is little demand on the system are often encouraged to do so by charging for off-peak energy at a reduced rate. This energy may be used for heating water which is stored in lagged tanks until required, or it may be used for heating night storage heaters which are arranged to provide central heating during the daytime. The aim of the supply authority is to raise the average daily load and to keep the generating and distribution plant constantly in use and earning revenue.

Electric welding may be carried out with reactors used to control the current supply but this results in the consumer operating at a very low power factor. It is to balance this demand for reactive voltamperes that the first tariff includes a penalty clause for welding equipment.

Summary

By improving the power factor at which a consumer operates his current demand on the system is reduced. Equipment such as capacitors or synchronous motors operating at a leading power factor may be installed to take leading reactive voltamperes which neutralise the lagging reactive load voltamperes.

Electricity tariffs are designed to ensure that the fixed and running costs of supply are covered. Penalties may be incurred by consumers operating at low power factors.

Examples 5.3.

1. What is meant by the power factor of an a.c. circuit? What are the disadvantages of a low power factor and how may this be improved?

A factory has a number of single-phase a.c. motors connected to a 400 V 50 Hz supply. The power demand on full load totals 10 kW at a power factor of 0·7 lagging. Calculate the value of a capacitor necessary to improve the power factor to 0·9 lagging when the machines are on full load. Determine also the supply current before and after the improvement. Sketch a phasor diagram.

2. Why is a.c. plant rated in kVA?

At full load and at 0·74 power factor lagging the efficiency of a 240 V single-phase motor of output 25 kW is 87%. Calculate the supply current at full load under these conditions and also when power factor correction is applied at the terminals of the motor so that the voltage and current are in phase.

3. How may the earning capacity of a supply system be reduced when consumers operate at a low power factor?

The current in an inductive circuit is 25 A and the power dissipated is 2 kW when the applied p.d. is 200 V at 50 Hz. Deduce the value of capacitor required to reduce the supply current to half the former value.

4. Explain, with the aid of a phasor diagram and a diagram of connections, how the operating power factor of a single-phase motor may be

raised from 0·4 to 0·9 lagging. Calculate the value of the capacitor required for the above improvement if the supply is 240 volts at 50 Hz and the current taken by the motor is 18 amperes prior to correction.

5. A small factory has a 240 volt 50 Hz single-phase supply. The lighting loads amount to 4·8 kW and may be assumed to be purely resistive. The remaining load comprises twelve similar single-phase motors with equal loads. The total current taken by these motors amounts to 120 amperes at 0·5 lagging power factor. Calculate the total kW and kVA loading of the factory, the total supply current and also the active and reactive components of the supply current.

What value of capacitance connected across the consumer's supply terminals would limit the total current to 80 amperes? What would be the magnitude and phase of the total current if six machines were switched off, with the correction capacitor remaining in circuit?

6. How could a three-phase synchronous motor be used to improve the power factor of a distribution system?

The efficiency of a 415 volt 50 Hz three-phase motor with an output of 75 kW when operating on full load at a power factor of 0·72 lagging is 86%. Determine the capacitance per phase of a mesh-connected capacitor to raise the power factor to unity. Include a diagram showing the capacitor connections.

7. Why are electricity tariffs commonly based on a two-part system?

A consumer has a maximum load of 210 kW at 0·75 power factor and an annual consumption of 480 000 kWh. The tariff is £5 per kVA of maximum demand plus 0·55 p per kWh. Calculate (a) the average cost per kWh over a period of one year and (b) the annual saving if the power factor is corrected to unity.

8. A consumer of electrical energy pays according to the following tariff: £6 per annum per kVA of maximum demand plus 1·15 p per kWh.

Calculate the average overall cost per kWh in a year when the annual consumption is 600 000 kWh with a maximum demand of 400 kW at 0·8 power factor. Describe briefly a suitable method of reducing the overall cost per kWh.

9. What are the main costs of supply which must be considered when compiling tariffs to be charged to consumers? What are the disadvantages to the supply authority of consumers who operate with low power factors? Compare the annual cost to a consumer of the following alternative tariffs when his maximum demand is 140 kW at 0·75 power factor and the annual consumption is 500 000 kWh.

Tariff A £6 per kVA of maximum demand + 0·5 p per kWh
Tariff B 0·7 p per kWh.

At what total annual consumption would the charges be the same under both tariffs?

6. Transformers

Introduction

Transformers are used for changing power in alternating current systems from one voltage to another. Let us look at one instance where it may be desirable to make such changes.

Consider a single-phase transmission line with a resistance of R ohm per metre. When a current of 5 amperes is delivered by the line at 1000 volts the power transmitted is 5 kW and the I^2R power loss amounts to $25R$ watts. If the voltage at the sending end of the transmission line is increased to 10 000 volts then the power transmitted by the same current and with equal losses will be 50 kW. Thus a given transmission line will transmit more power with the same losses if a higher transmission voltage is used. Hence the efficiency of the transmission is improved.

By using a static transformer at the generating station an alternating voltage may be readily transformed to a higher value prior to transmission. At the receiving end of the transmission line the voltage is reduced by a second transformer and power is fed to the distributing networks. Transformers operate with a very high efficiency. They have no moving parts and require very little maintenance.

When dealing with electromagnetic induction in Volume 1, Chapter 5 we saw how Faraday in an experiment with two coils wound on an iron ring demonstrated induced e.m.f.'s. A change in current through one coil produced a change in magnetic flux linking with the other coil and this in turn produced an induced e.m.f. This experiment illustrated the principle of the transformer. If instead of opening and closing a switch in the primary circuit as Faraday did we apply an alternating voltage to this circuit then the resulting changing flux will induce alternating e.m.f.'s in the windings with which it links.

The transformer on no load

Here is an experiment to demonstrate the action of a transformer on no load.

Experiment 6(i) *To vary the voltage applied to the primary winding of a transformer and to observe the no-load current, the input power and the voltages which appear across two secondary windings*

The circuit was connected as in Fig. 6.1.

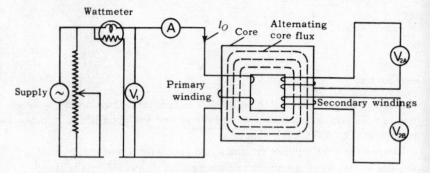

Fig. 6.1. Single-phase transformer on no load.

The primary and two secondary windings were wound on a laminated iron core which provided a low reluctance path for the alternating flux linking with the coils.

The voltage applied to the primary winding was varied in steps of 20 volts from 50 volts to 270 volts. Corresponding values of input power, current and the voltages appearing across the two secondary windings were observed and tabulated.

V_1	(V)	50	70	90	110	130	150	170	190	210	230	250	270
V_{2A}	(V)	10	13	16	18	23	25	29·5	32	35	39	42	45
V_{2B}	(V)	24	33	43	53	62	72	81	91	101	110	120	130
I_0	(A)	0·07	0·1	0·12	0·15	0·18	0·21	0·25	0·31	0·38	0·5	0·68	0·92
P	(W)	1·6	3·6	5·9	9·0	12·9	16·6	20·3	24·7	29·4	34	38	44
$\cos \phi$		0·46	0·51	0·55	0·55	0·55	0·53	0·48	0·42	0·37	0·3	0·22	0·18
$\sin \phi$		0·89	0·86	0·83	0·83	0·83	0·85	0·88	0·91	0·92	0·95	0·97	0·98
$I_P = I_0 \cos \phi$	(A)	0·032	0·05	0·066	0·08	0·01	0·11	0·12	0·13	0·14	0·15	0·15	0·16
$I_M = I_0 \sin \phi$	(A)	0·062	0·86	0·10	0·12	0·15	0·18	0·22	0·28	0·35	0·48	0·66	0·9

Primary windings N_1 — 192 turns.   Secondary windings: N_{2A} — 32 turns
N_{2B} — 92 turns

The primary and secondary turns and voltage ratios

Let us look at the primary and secondary winding turns and voltage ratios.

The voltages appearing across the secondary windings bear a constant relationship to the primary applied voltage. Taking mean values

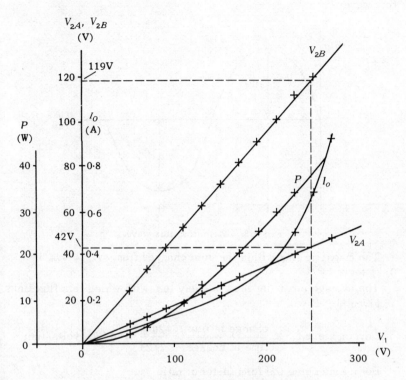

Fig. 6.2. Transformer no-load characteristics.

from the curves,

$$\frac{V_1}{V_{2A}} = \frac{250}{42} = 6 \text{ and } \frac{V_1}{V_{2B}} = \frac{250}{119} = 2\cdot1$$

Now, compare these ratios with the turns ratios,

$$\frac{N_1}{N_{2A}} = \frac{192}{32} = 6 \text{ and } \frac{N_1}{N_{2B}} = \frac{192}{92} = 2\cdot1$$

It is thus seen that the ratio of the voltages across two windings in a transformer, is equal to the turns ratio of these windings. How can we explain this?

Deriving the e.m.f. equation for a transformer

Let us assume that the primary applied alternating e.m.f. of frequency f(Hz) has a sinusoidal waveform and that the resulting flux has a similar type of waveform. Figure 6.3 shows one cycle of the magnetic flux wave.

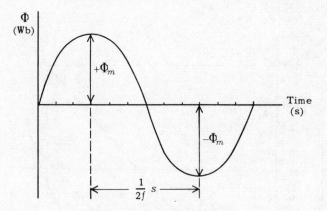

Fig. 6.3. Alternating flux wave.

The magnetic flux within the core changes from $+I_m$ to $-I_m$ in $1/2f$ second.

The average e.m.f. induced in every turn with which this flux links is given by,

$$\text{average e.m.f.} = \frac{\text{change in flux}}{\text{time of change}} = \frac{2\Phi_m}{1/2f} = 4f\Phi_m \text{ volts/turn}$$

For a sine curve the form factor or ratio

$$\frac{\text{r.m.s.}}{\text{average}} = 1\cdot 11$$

Thus the r.m.s. value of the induced e.m.f. $= 4\cdot44 f\Phi_m$ volts/turn
If a secondary winding has N_{2A} turns the induced e.m.f.

$$E_{2A} = 4\cdot44 f\, \Phi_m N_{2A} \text{ volts}$$

Similarly for a winding with N_{2B} turns the induced e.m.f.

$$E_{2B} = 4\cdot44 f\, \Phi_m N_{2B} \text{ volts.}$$

The e.m.f. induced in the primary winding will be a back e.m.f. opposing the applied voltage and the magnitude is given by,

$$E_1 = 4\cdot44 f\, \Phi_m N_1 \text{ volts.}$$

Therefore,

$$\frac{E_1}{E_{2A}} = \frac{N_1}{N_{2A}} \text{ and } \frac{E_1}{E_{2B}} = \frac{N_1}{N_{2B}}$$

When a transformer is on no load the currents in the windings are very low and if we neglect the internal winding voltage drops we may write,

$$\frac{V_1}{V_{2A}} = \frac{N_1}{N_{2A}} \text{ and } \frac{V_1}{N_{2B}} = \frac{N_1}{N_{2B}}$$

These ratios are confirmed by the result of the preceding experiment.

An alternative method of deriving the e.m.f. equations is as follows:

let ϕ = instantaneous value of the core flux.

Assuming sinusoidal waves,

$$\phi = \Phi_m \sin 2\pi f t$$

The e.m.f. induced in a single turn,

$$e = -\frac{d\phi}{dt} = -\frac{d}{dt}(\Phi_m \sin 2\pi f t)$$
$$= -2\pi f \, \Phi_m \cos 2\pi f t \text{ volts per turn}$$
$$= 2\pi f \, \Phi_m \sin(2\pi f t - \pi/2)$$

This indicates that the induced e.m.f. wave lags $\pi/2$ radians behind the flux wave (Fig. 6.4).

$$\text{Maximum e.m.f.} = 2\pi f \, \Phi_m \text{ volts per turn}$$

$$\text{R.M.S. e.m.f.} = \frac{2\pi f}{\sqrt{2}} \Phi_m \text{ volts per turn}$$

Hence,

$$E_1 = 4 \cdot 44 f \Phi_m N_1 \text{ volts}$$
$$E_{2A} = 4 \cdot 44 f \Phi_m N_{2A} \text{ volts}$$
$$E_{2B} = 4 \cdot 44 f \Phi_m N_{2B} \text{ volts}.$$

V_1 will be equal and opposite to E_1 if we neglect any impedance drops within the windings.

The transformer phasor diagram on no load

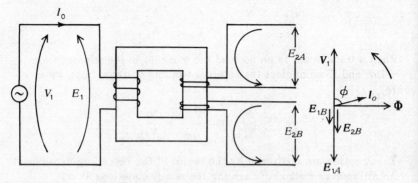

Fig. 6.4. Transformer no-load phasor diagram.

Figure 6.4 shows the circuit and phasor diagrams for the transformer with 250 volts applied across the primary winding.

Induced e.m.f.'s lag behind the flux wave by $\pi/2$ radians and as internal voltage drops resulting from winding impedances are being neglected the voltage (V_1) applied to the primary winding is shown equal and opposite to the e.m.f. (E_1) induced in this winding.

From the readings of the wattmeter, ammeter and voltmeter in the preceding experiment we may determine the position of the no load current phasor (I_0).

With 250 volts applied to the primary winding,

$$\cos \phi = \frac{\text{watts}}{\text{voltamperes}} = \frac{38}{250 \times 0 \cdot 68} = 0 \cdot 224$$

If there were no power losses in the transformer on no load then the current taken from the supply would be a purely magnetising current lagging $\pi/2$ radians behind the supply voltage. The primary winding would behave in the same manner as a pure inductor connected across an alternating voltage supply.

No-load losses in a transformer

With no load on the secondary side there is an $I_0^2 R$ power loss in the primary winding but because of the low winding resistance and small no-load primary current this loss will be neglected.

When the transformer is on no load, with normal voltage applied to the primary winding, the full flux is present in the core and the no-load iron losses will be the same as when the transformer is on load. These losses must be taken into account.

To supply the hysteresis and eddy-current core-losses (see page 230) the magnetising current has an active or power component $I_0 \cos \phi$ in phase with the supply voltage.

The wattless or reactive magnetising component, $I_0 \sin \phi$ lags behind the applied voltage by $\pi/2$ radians.

The curves (Fig. 6.5(a)) show how the relative magnitudes of $I_0 \cos \phi$ and $I_0 \sin \phi$ vary as the primary voltage is increased.

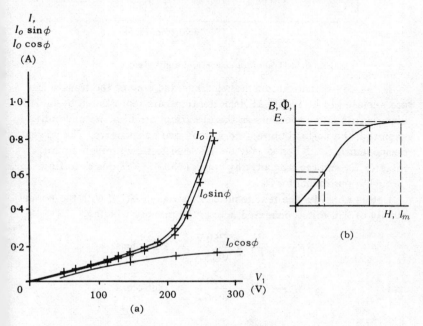

Fig. 6.5. Variation of transformer no-load current components with applied voltage.

At the higher values of applied voltage the magnetising component $I_0 \sin \phi$ increases rapidly. This is because of the levelling of the $B - H$ curve for the material of the magnetic core. As B increases (Fig. 6.5(b)), comparatively larger increases in magnetising current are required to bring about small increases in the core flux and the induced e.m.f.

Representing a transformer on no load by an equivalent circuit

The power component $I_0 \cos \phi$ of the no-load current rises as V_1 is increased. On no load the transformer windings and core may be represented by an equivalent circuit as shown in Fig. 6.6.

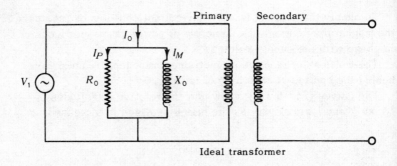

Fig. 6.6. Transformer no-load equivalent circuit.

In the equivalent circuit the windings and core of the transformer are assumed to be ideal. All imperfections are represented by includding additional components in the electrical circuits. We have already seen that the no-load current comprises two components. The power component $I_P = I_0 \cos \phi$ may be assumed to flow through a pure resistor R_0 and the magnetising component $I_M = I_0 \cos \phi$ to flow through a pure inductor X_0.

Let us consider the test results (Experiment 6(i)), with the normal supply of 250 volts connected across the primary winding,

$$R_0 = \frac{V_1}{I_P} = \frac{250\,\text{V}}{0\cdot15\,\text{A}} = 1733\,\text{ohms}$$

and
$$X_0 = \frac{V_1}{I_M} = \frac{250\,\text{V}}{0\cdot66\,\text{A}} = 380\,\text{ohms}$$

The transformer as an example of mutual inductance

We have already dealt with self inductance in Volume 1 and now that we have carried out a no-load test on a transformer let us have a look at the transformer as an example of mutual inductance.

If the current in coil A (Fig. 6.7) changes then this will be accompanied by a change in flux linking with coil B and an e.m.f. is induced in B. The coils are said to possess mutual inductance.

The unit of mutual inductance (M) is the henry. Two coils have a mutual inductance of one henry if an e.m.f. of one volt is induced in one coil as the current in the other coil changes at the rate of one ampere per second.

Consider two coils A and B with mutual inductance M henrys. Let the current in coil A change from i_1 to i_2 in t seconds whilst the accompanying flux through coil B changes from Φ_1 to Φ_2.

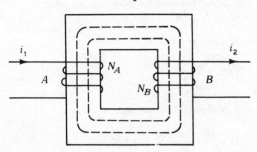

Fig. 6.7. Mutual inductance between two coils.

The e.m.f. induced in coil $B = -M(i_2 - i_1)/t$ volts. We already know however that the e.m.f. induced in coil B

$$= -\frac{(\Phi_2 - \Phi_1)}{t} \times N_B \text{ volts}$$

The negative sign indicates that the induced e.m.f. tends to establish a current which opposes any change in flux linking with the coils.

Hence $M = \dfrac{(\Phi_2 - \Phi_1)}{(i_2 - i_1)} \times N_B$ henrys

$= \dfrac{\text{change in flux linkages in the secondary circuit}}{\text{change in primary current}}$

= flux linkages in coil B per ampere in coil A.

Example

If coils A and B (Fig. 6.7) have 4 and 6 turns respectively and if 1 ampere in coil A produces a core flux of 2 Wb, calculate the self-inductance of both coils and the mutual inductance between them. Assume that all the flux produced by one coil links with the other coil, and that there is no saturation of the magnetic circuit.

$$L_A = \frac{\Phi_A \times N_A}{I_A} = \frac{2 \times 4}{1} = 8\,\text{H}$$

If one ampere flows in coil B it will produce more ampere turns than one ampere flowing in coil A. Since there is no saturation of the magnetic circuit, flux produced by one ampere in coil
$$B = 2 \times 6/4 = 3\,\text{Wb}$$

$$L_B = \frac{\Phi_B \times N_B}{I_B} = \frac{3 \times 6}{1} = 18\,\text{H}$$

Consider one ampere in coil A

$$M = \frac{\Phi_B \times N_B}{I_B} = \frac{2 \times 6}{1} = 12\,\text{H}$$

Neglecting any leakage flux we may write,

$$M = \frac{\Phi_B \times N_B}{I_A} \quad \text{and} \quad M = \frac{\Phi_A \times N_A}{I_B}$$

$$\therefore M^2 = \frac{\Phi_B \times N_B}{I_B} \times \frac{\Phi_A \times N_A}{I_A} = L_A L_B$$

$$\therefore M = \sqrt{L_A L_B}$$

If the whole of the flux produced by one coil does not link with the other coil then this relationship may be expressed as

$$M = k\sqrt{L_A L_B} \quad \text{and} \quad k = \frac{M}{\sqrt{L_A L_B}}$$

where k is termed the coefficient of coupling.

Mutual and self inductances of two coils connected in series

Let us consider two series connected coils which have self inductance L_1 and L_2 and also mutual inductance M. (Fig. 6.8)

A change in the current through coil 1 in Fig. 6.8(a) will result in,

(a) an e.m.f. being self induced in coil 1 and
(b) an e.m.f. being mutually induced in coil 2.

Since the coils are in series the current will also change in coil 2 and this is accompanied by,

(a) an e.m.f. being self induced in coil 2 and
(b) an e.m.f. being mutually induced in coil 1.

You have already seen in Chapter 3 how Kirchhoff's second law may be applied to a.c. circuits. An inductor was represented by $j\omega L$ and the e.m.f. self induced when an alternating current I flows through it is given by $I \times j\omega L$. Now mutual inductance M also results in induced e.m.f.'s. The direction of the mutually induced e.m.f. depends on whether the linking flux produced by M aids or opposes that produced by the self inductance. If the flux aids then the e.m.f. induced may

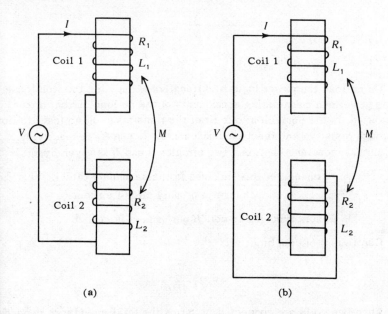

Fig. 6.8. Self and mutual inductance of coils in series.

be expressed in the j notation by $j\omega M \times I$ and if it opposes by $-j\omega M \times I$.

When two coils are wound so that the magnetic fields aid each other as in Fig. 6.8(a) then applying Kirchhoff's second law to this circuit we may write,

$$\mathbf{V} = R_1 I + j\omega L_1 I + j\omega MI + R_2 I + j\omega L_2 I + j\omega MI$$

$$\mathbf{Z} = \frac{\mathbf{V}}{I} = (R_1 + R_2) + j\omega(L_1 + L_2 + 2M)$$

The effective inductance in the circuit is $(L_1 + L_2 + 2M)$ henrys.

If the coils are wound as in Fig. 6.8(b) so that the magnetic fields oppose each other then the effective inductance in the circuit is given by $(L_1 + L_2 - 2M)$ henrys.

Summary

Neglecting losses in a transformer the voltage and turns ratios are related by the expressions,

$$\frac{V_1}{V_2} = \frac{N_1}{N_2}$$

Induced e.m.f. = $4 \cdot 44 \, f \, \Phi_m \, N$ volts.

On no load the power input to a transformer supplies the iron losses. The power and magnetising components of the no load current are assumed, in the equivalent circuit of the transformer on no load, to flow through resistive and reactive components, R_0 and X_0 respectively.

Mutual inductance between two circuits A and B is given by,

$$M = \frac{\text{change in flux linkages in the secondary circuit}}{\text{change in primary circuit current}}$$

M = flux linkages in coil B per ampere in coil A.

Coefficient of coupling,

$$k = \frac{M}{\sqrt{L_A L_B}}$$

When two coils are connected in series the total inductance depends on whether the coils are aiding or opposing and is given by,

$$(L_A + L_B \pm 2M) \text{ henrys.}$$

Examples 6.1.

1. Describe the principle of action of a transformer. Sketch the diagram of a circuit which could be used in a test to determine the relationship between the primary and secondary no-load terminal voltages of a transformer.

A transformer is designed to operate on a 240 V 50 Hz supply. The secondary windings are required to provide 6 volts and 2 volts respectively. Determine the turns ratio between the primary and secondary windings. Would it be possible to obtain a 4 volt output from the above windings?

2. A step-down transformer has a turns ratio of 15 : 1 and a primary voltage of 3300 volts is applied. Calculate the value of the e.m.f. induced in the secondary winding. If the full load output is 25 kVA calculate the secondary current on full load when the load power factor is unity. Neglect all losses.

3. Derive an expression for the e.m.f. induced in a transformer winding in terms of the supply frequency, the maximum value of the core flux and the number of turns in the winding.

A 10 kVA single-phase transformer has a turns ratio of 900/30. The primary is connected to a 250 volt 50 Hz supply. Calculate the open circuit secondary volts and the maximum value of the core flux. Neglect all losses.

4. A single-phase transformer is required to give an output of 10 amperes at 12·5 volts when the primary is connected across a 240 volt 50 Hz supply. If the cross section of the core is 250 mm² and the maximum flux density is 0·8 tesla determine (a) the primary turns and (b) the secondary turns. Neglect all losses in the calculations.

5. The core of a transformer has an iron cross-sectional area of 0·02 m² and is wound with coils of 400 and 1000 turns which constitute respectively the primary and secondary windings. Determine the maximum value of the flux-density in the core and the induced voltage in the secondary winding when the primary winding is connected to a 440 volt 50 Hz supply. Derive any formula used.

6. A single phase transformer has 11 000 volts applied to the primary winding of 2000 turns and is required to supply power to equipment with a maximum load of 10 kW at 400 volts and 0·8 lagging power factor. Estimate the number of secondary turns required and the secondary current on full load. Neglect all losses.

7. What happens to the power input of a transformer when the secondary winding is on no-load?
Determine the component iron loss currents in a transformer if the input power is 75 watts at 240 volts and the power factor is 0·25 on no load. Sketch a phasor diagram representing the supply voltage, induced e.m.f. and current.

8. Describe the various losses which occur in a transformer on no load. How may a transformer on no load be represented by an equivalent circuit?
In a no-load test on a 200 volt single-phase transformer the input current was 0·23 amperes and the input power 24 watts with the normal primary voltage applied. Calculate the values for the iron loss components in the equivalent circuit. Sketch a phasor diagram and show (a) active and reactive components of the no-load current (b) the applied voltage and (c) the induced e.m.f.

9. A single-phase transformer is required to develop 1100 volts across the secondary winding on no load, when the primary is connected to a 240 V 50 Hz supply. Estimate the number of turns in each winding if the maximum core flux-density is 0·8 T and the core cross-sectional area is 0·008 m².

10. Explain the meanings of self inductance and mutual inductance. Two identical coils each have 800 turns and are wound on non-magnetic formers. A current of 2 A through one of the coils produces a magnetic flux of 420 μWb. Calculate (a) the inductance of each coil (b) the average value of the e.m.f. induced in each coil when the current of 2 A is reversed in 0·5 second and (c) the mutual inductance between the coils if an e.m.f. of 1 V is induced in one coil when a current of 2 A is reversed in 0·5 second in the other coil.

11. Describe briefly two practical applications of (a) self-inductance and (b) mutual-inductance.

Two identical coils are wound co-axially a short distance apart. When one coil having a resistance of 45 ohms is connected to a 200 V 50 Hz supply, the current taken is 3 amperes and the e.m.f. induced in the second coil on open circuit is 100 volts. Calculate the self-inductance of each coil and the mutual inductance between them.

12. Discuss the factors upon which the inductance of a coil depends. How does introducing a core of magnetic material into the centre of two co-axial coils affect the mutual inductance?

Two coils A and B of 800 turns and 500 turns respectively are positioned so that all the flux produced by coil A links with coil B. If the self inductance of coil A is 2 mH determine (a) the mutual inductance between the coils and (b) the e.m.f. induced in coil B if the current in coil A changes at 500 A/s.

13. Explain the terms self-inductance and mutual-inductance defining the units in which each is measured.

Two air-cored coils, A of 600 turns and B of 900 turns, are mounted side by side so that 50% of the flux produced by A links with B. If a current of 2 A in coil A produces a flux of 300μWb determine (a) the self-inductance of coil A, (b) the mutual-inductance between coils A and B and (c) the total inductance if the coils are connected in series aiding.

14. Explain the meaning of self-inductance and mutual inductance. A coil having a self inductance of 0·01 H and a resistance of 10 ohms carries a current which alternates between +10 A and −10 A at the rate of 1000 A/s as a triangular waveform. Determine and plot to scale, waveforms of current and applied voltage over one cycle, and evaluate the frequency.

15. Two air-cored coils are disposed relatively to each other so that 25% of the flux in one coil links with all the turns of the other coil. Each coil has a mean cross section of 0·03 m^2 and a length of 1 metre. If there are 2000 turns of wire in one coil determine the number of

turns required in the other coil to give a mutual inductance of 4·6 mH.

16. Two separate inductors of 50 mH and 150 mH respectively, are mounted so that there is a mutual inductance of 40 mH between them. What will be the resulting inductance between them when the two coils are connected (a) in series aiding and (b) in series opposing?

17. How is the coefficient of coupling related to the self inductances and mutual inductance of two adjacent coils? Two coils connected in series aiding have self inductances of 24 mH and 32 mH respectively. If the total inductance is 84 mH determine the coefficient of coupling. What would be the total inductance if the coils were in series and opposing?

Load test on a single-phase transformer

Now that you have studied a transformer on no load let us carry out a test on a single phase transformer supplying a load which is largely resistive.

Experiment 6(ii) *To carry out a load test on a transformer*

The transformer was connected as shown (Fig. 6.9).

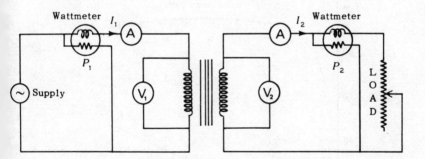

Fig. 6.9. Transformer load test circuit diagram.

The primary winding was connected to a 230 volt 50 Hz supply and the secondary load which consisted of a variable resistance unit was adjusted to steps until the increasing secondary current reached full load. All instrument readings were observed at each stage and tabulated.

V_1 (V)	V_2 (V)	I_1 (A)	I_2 (A)	P_1 (W)	P_2 (W)	$\cos\phi_1$ $P_1/V_1 I_1$	$\cos\phi_2$ $P_2/V_2 I_2$	Efficiency (%)
230	118	0·13	–	12·6	–	0·42	–	–
230	118	1·12	2	238	218	0·92	0·93	90·8
230	117	1·64	3	360	338	0·05	0·97	93·8
230	116	2·11	4	460	440	0·95	0·95	96·0
230	114	2·59	5	571	548	0·96	0·96	96·0
230	113	3·19	6	708	662	0·96	0·98	93·5
230	112	3·69	7	816	758	0·96	0·97	92·8
230	111	4·17	8	930	860	0·97	0·97	92·5
230	110	4·65	9	1028	954	0·96	0·97	92·8

Transformer rating: 1 kVA
Primary turns: $N_1 = 142$. Secondary turns: $N_2 = 72$.

The transformer load characteristics were plotted as in Fig. 6.10.

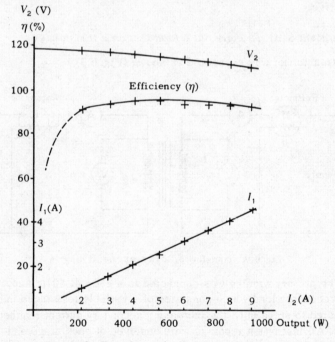

Fig. 6.10. Transformer load characteristics.

The turns ratio

$$\frac{N_2}{N_1} = \frac{72}{142} = 0\cdot51$$

From the experimental results it is observed that the mean current ratio,

$$\frac{I_1}{I_2} = \frac{4 \cdot 6}{9} = 0 \cdot 51$$

and the voltage ratio on no load,

$$\frac{V_2}{V_1} = \frac{118}{230} = 0 \cdot 51$$

It is evident that the efficiency of even this small transformer reaches a high value. Power losses which occur are the I^2R losses in the windings and the magnetic-hysteresis and eddy-current core losses.

If we assume 100% efficiency on full load then theoretically,

$$\text{input power} = \text{output power}$$
$$V_1 I_1 \cos \phi_1 = V_2 I_2 \cos \phi_2$$

and since the power factor is almost the same on both sides of the transformer we may write,

$$V_1 I_1 \simeq V_2 I_2$$
$$\frac{I_1}{I_2} \simeq \frac{V_2}{V_1}$$

This expression may be extended and written as

$$\frac{I_1}{I_2} \simeq \frac{V_2}{V_1} \simeq \frac{N_2}{N_1}$$

The experimental results are seen to confirm the equality of these ratios.

From this last expression we may write $I_1 N_1 \simeq I_2 N_2$ showing that the primary and secondary ampere turns balance.

When a transformer is on no load the magnetising current establishes an alternating flux until the e.m.f. induced in the primary winding is almost equal and opposite to the primary applied voltage. If load current flows in the secondary winding then this winding establishes a demagnetising flux which according to Lenz's law tends to reduce the main core flux. This in turn reduces the back e.m.f. induced in the primary winding and allows more current to flow from the supply. The result is that the increase in primary ampere turns counteracts the secondary demagnetising ampere turns.

Transformer phasor diagram on load, neglecting winding losses

Figure 6.11 is a phasor diagram representing the transformer used in **Experiment 6(ii)** when the secondary winding is delivering a load current of 9 amperes.

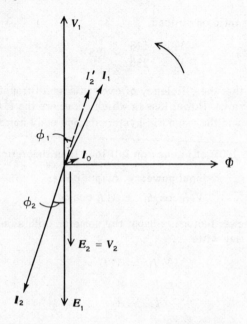

Fig. 6.11. Phasor diagram with resistive loading.

E_1 and E_2 represent the e.m.f.'s induced in the transformer windings and both e.m.f.'s lag by $\pi/2$ radians behind the core flux. The applied voltage V_1 is assumed equal and opposite to E_1. It is seen from the the test results that the secondary current lags slightly behind V_2 since the load is not of pure resistance. Hence we may draw I_2 displaced by ϕ_2 radians from V_2 where $\cos \phi_2 = 0.96$.

A current must flow in the primary winding to supply the iron losses and produce magnetic flux in the core of the transformer. This no-load current is shown by I_0. It lags behind V_1 by an angle ϕ_1 such that $\cos \phi_1 = 0.42$ as determined from the test results when the load current is zero.

Current must also flow in the primary winding to neutralise the ampere turns of the secondary winding when it carries load current. This primary current is represented by I_0'. The magnitude is equal to $I_2 \times N_1/N_2$. I_2' may be termed 'the secondary load current referred to the primary circuit'.

The total primary current will be given by

$$I_1 = I_0 + I_2'$$

lagging by the angle ϕ_1 behind the supply voltage.

Transformer winding resistances and leakage reactances

Until now we have neglected any resistance in the transformer windings. In practice resistance is always present and this produces internal voltage drops and power losses in the windings. To keep these losses to a low value windings are designed to have low resistance,

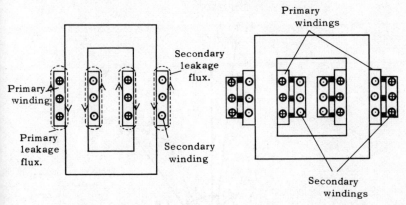

(a) Coils on separate limbs.   (b) Core type with concentric coils.

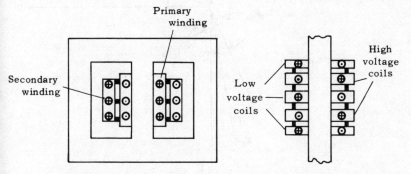

(c) Shell type with concentric coils.   (d) Sandwich coil arrangement.

Fig. 6.12. Physical arrangement of transformer windings.

211

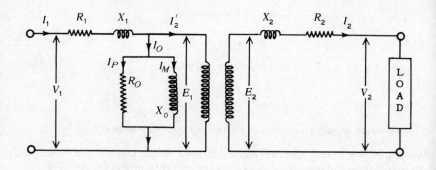

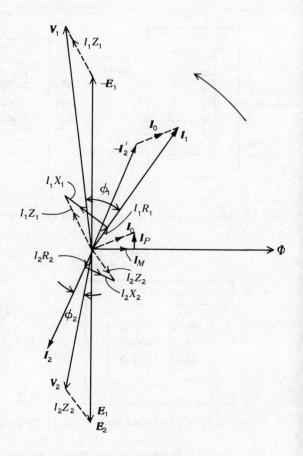

Fig. 6.13. Transformer equivalent circuit and phasor diagram.

There is inevitably some leakage of magnetic flux in a transformer. All the flux produced by one winding may not link with the other windings in the transformer. Leakage conditions are worst when the primary and secondary windings are on separate pole limbs (Fig. 6.12 (a)). This was the type of transformer tested in **Experiment 6(ii)**.

In (b), (c) and (d) of Fig. 6.12 alternative forms of core and coil arrangements are shown. These designs help to reduce leakages fluxes. Leakage flux which links with only one winding produces a leakage reactance and this in turn causes an internal voltage drop in the winding with which it links.

Resistances (R) and leakage reactances (X) of windings may be represented by including appropriate components in the equivalent circuit of the transformer. Figure 6.13 shows an equivalent circuit and corresponding phasor diagram.

Φ represents the magnetic flux and E_1 and E_2 the e.m.f.'s induced in the primary and secondary windings. For simplicity, the transformer is assumed to have equal primary and secondary turns and hence E_1 and E_2 are equal. The secondary load current is I_2 and as the load is assumed to be slightly inductive I_2 lags behind E_2. Owing to the resistance and leakage reactance of the secondary winding, internal voltage drops $I_2 R_2$ and $I_2 X_2$ will be in phase and in quadrature respectively with I_2. The p.d. across the load will be less than E_2 and is represented by V_2. The power factor of the load is given by cos ϕ_2.

Corresponding to the current I_2 in the secondary winding there will be a current I'_2 in the primary winding. With equal turns on both windings these currents will be of equal magnitudes. The total primary current I_1 will be I'_2 combined with the no load current I_0 which has power and magnetising components I_P and I_M.

As a result of the primary winding resistance R_1 and leakage reactance X_1 internal voltage drops $I_1 R_1$ and $I_1 X_1$ will be in phase and in quadrature respectively with I_1. The applied voltage must provide these internal voltage drops and also neutralise the induced e.m.f. E_1. The required applied p.d. is therefore represented by V_1 and the power factor presented to the supply is cos ϕ_1.

Let us now consider capacitive load conditions.

Testing a transformer under capacitive load conditions

Experiment 6(iii) *To investigate the secondary terminal voltage of a transformer when the load is partially capacitive*

The same circuit was connected as in **Experiment 6(ii)** but the load consisted of a variable capacitor and resistor connected in series. The secondary voltage was first measured with a current of 4 amperes flowing through a resistive load. Capacitance was then introduced into the load circuit and the loading adjusted to maintain a constant load current of 4 amperes at various leading power factors. The observations were as tabulated.

I_2 (A)	P_2 (W)	V_2 (V)	P.F. $\cos \phi_2$
4	480	120	1·0
4	440	121	0·91
4	300	122	0·62
4	250	124	0·5
4	200	125	0·4
4	10	127	0·02

It is seen that (Fig. 6.14) the secondary terminal voltage tends to rise as the load current becomes more leading.

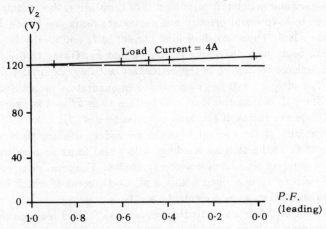

Fig. 6.14. How the secondary terminal voltage varies with constant load current at increasing leading power factors.

How can we account for this variation in voltage with power factor?

Let us refer to a general phasor diagram for a transformer supplying a load with a leading power factor (Fig. 6.15).

The secondary load current I_2 is shown leading the secondary voltage. As a result of the secondary winding resistance and leakage reactance we have a total secondary winding impedance drop given by $I_2 Z_2$. The secondary induced e.m.f. must be equal to the resultant of V_2 and $I_2 Z_2$. It will be appreciated that the greater the angle by

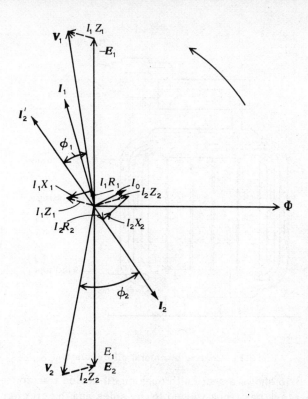

Fig. 6.15. Phasor diagram for a transformer with capacitive loading.

which I_2 leads V_2 then the greater will be the effect of $I_2 Z_2$ in increasing V_2. This is the reason for the rising voltage characteristic obtained in the experiment.

A current I_2' will flow in the primary circuit and this current combined with I_0 gives the total primary current I_1. The primary applied voltage must be equal to the resultant of $-E_1$ and the voltage drop $I_1 Z_1$.

The magnitudes of the magnetising current and voltage drops $I_1 Z_1$ and $I_2 Z_2$ are considerably exaggerated in the diagram (Fig. 6.15).

Cooling of transformers

The output of a transformer is limited by the maximum permissible temperature rise. The losses within the transformer which produce a rise in the operating temperature are the core iron losses and the winding $I^2 R$ losses.

Natural air cooling is limited to small transformers while power transformers are generally oil cooled.

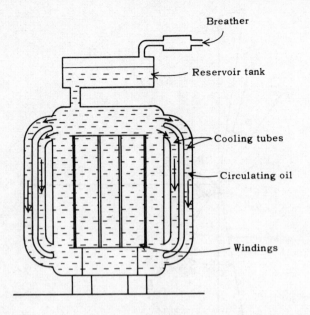

Fig. 6.16. Cooling by natural oil circulation.

Figure 6.16 shows a section through an oil-cooled transformer. The transformer tank has tubes welded to the sides and the oil circulates through these tubes. Warm oil rises to the top of the tank, flows down the tubes and as it cools enters the tank near the base. A reservoir tank is provided to maintain the oil level as the oil in the main tank expands and contracts with temperature changes. A breather fitted to the reservoir ensures that any air which enters when the oil volume contracts is dehydrated. The power dissipated by a tubular tank may be increased by using a fan to produce an air blast across the cooling tubes. In another system oil is pumped through cooling radiators against which a forced air blast is directed.

Summary

Neglecting losses, the primary and secondary winding ampere-turns in a transformer must balance.

$$\frac{I_1}{I_2} = \frac{N_2}{N_1} = \frac{V_2}{V_1}$$

Winding resistance and leakage fluxes may be represented by including resistances and reactances in the transformer equivalent circuit

With capacitive loading the output voltage may rise slightly with increasing load current.

Examples 6.2.

1. Describe how you would proceed to carry out a load test on a transformer. Include a circuit diagram and sketch typical curves showing secondary volts, efficiency and secondary current plotted against output load.

2. An a.c. load of 0·8 power factor takes 2 amperes at 100 volts and a 200 volt a.c. supply is available. Explain how the 100 volt supply may be obtained by using a transformer and calculate the turns ratio. What value of series resistor could be used instead of the transformer? Which method would you expect to give the better overall efficiency and why?

3. Explain the action of a single phase transformer and derive an expression relating primary and secondary currents, voltages and turns. Neglect all losses.

A 2 kVA single-phase transformer is connected to a 240 volt 50 Hz supply and has 1400 turns on the primary winding. How many turns must be provided on secondary windings to supply loads at 60 volts and 80 volts? Calculate the full load primary current at unity power factor.

4. Explain how when a transformer is loaded the primary and secondary ampere turns almost balance each other. A 20 kVA single-phase step down transformer has a turns ratio of 300/23 and the primary winding is connected to a 1500 volt 50 Hz supply. Calculate the primary and secondary voltages and currents on full load, neglecting all losses.

5. What determines the primary current of a transformer (a) on no load and (b) when it is loaded?

A single-phase transformer has a no-load voltage ratio of 415/240 volts, and takes a no-load current of 3 amperes at 0·15 power factor. Determine graphically or otherwise the magnitude of the primary current when the secondary load is 15 amperes at 0·8 lagging. Neglect winding resistances and assume that the same flux passes through both windings.

6. Draw and explain a simple phasor diagram for a transformer supplying a lagging load. Neglect winding resistances and leakage reactances.

A 440/220 volt single-phase transformer takes a no-load current of 1·0 ampere at a power factor of 0·2. Determine the primary current, and the power factor when the secondary current is 24 amperes at 0·8 power factor.

7. Sketch an equivalent circuit for a transformer and include components to represent the core losses and also the primary and secondary winding resistances and leakage reactances.

A transformer is connected with the rated voltage across the primary winding. If it delivers one half of its rated full-load current output explain the effects of the components in your equivalent circuit when the load is increased until full-load current flows in the windings.

8. Distinguish between shell and core type single phase transformers illustrating your answer with sketches.

What is meant by leakage reactance when referred to the primary and secondary windings of a transformer?

Sketch a phasor diagram for a single-phase transformer supplying a secondary load of unity power factor. Explain clearly how the position of each phasor is determined. Take into account voltage drops caused by the primary and secondary winding resistances and leakage reactances. Include also the iron magnetising and power-loss components of the no-load current.

9. Explain why the voltage across the secondary terminals of a loaded transformer falls more when the load is inductive than when it is of pure resistance. How do the power factors compare on the primary and secondary sides of a transformer with resistive or partially inductive loading? Sketch phasor diagrams as may be necessary to illustrate your answer.

10. Sketch and explain carefully the phasor diagram for a transformer with a partially capacitive load. How would you expect the secondary voltage of a transformer to vary with increasing load if the load were (a) resistive and (b) highly capacitive?

11. Describe briefly two methods of cooling transformers. Explain why different methods are used for small and large transformers.

A 20 kVA single-phase transformer has the following losses at full-load unity power factor: iron losses 400 W, copper losses 500 W. Calculate the efficiency of the transformer at unity power factor (a) on full load and (b) on one half full load.

Simplified transformer equivalent circuits neglecting iron losses

On full load the effects of R_0 and X_0 are only very small and hence these components may be omitted from the equivalent circuit leaving us with a simplified equivalent circuit as in Fig. 6.17.

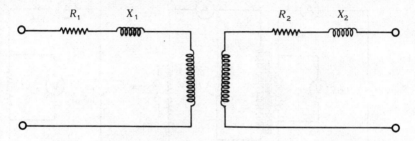

Fig. 6.17. Transformer equivalent circuit neglecting iron losses.

As a further simplification the separate winding impedances may be assumed to be in one circuit. Figure 6.18 shows these impedances included in the primary circuit. R_2' and X_2' signify respectively the resistance and reactance of the secondary winding referred to the primary circuit.

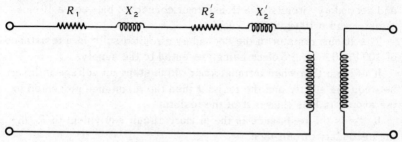

Fig. 6.18. Transformer equivalent circuit with winding impedances included in the primary circuit.

The transfer of resistance, reactance and impedance from one side of a transformer to the other

We have seen how the equivalent circuit of a transformer may be simplified by transferring the secondary winding resistance and leakage reactance to the primary side. If a transformer has equal turns in the primary and secondary windings this is a simple direct transfer of perhaps 0·5 ohm from one winding to the other. Let us carry out experiments to investigate the effect when the turns ratio is other than unity.

Experiment 6(iv) *To determine the equivalent resistance presented to the supply when a resistor is connected in the secondary circuit of a transformer*

(a) Transformer stepping up the voltage between a supply and resistor.

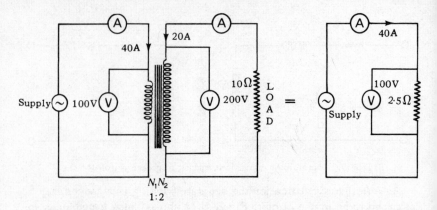

Fig. 6.19. Transferring resistance from the secondary to the primary circuit of a transformer.

Figure 6.19 shows the connections and conditions in the primary and secondary circuits of a transformer connected between a 10 ohm resistor and a 100 volt a.c. supply.

The 10 ohm resistor in the secondary circuit results in a resistance of 100 V/40 A = 2·5 ohms being presented to the supply.

It is clear that when a transformer which steps up voltage is inserted between the supply and the resistor then the resistance presented to the supply is less than that of the resistor.

If R_2' is the resistance in the primary circuit equivalent to R_2 in the secondary circuit then

$$R_2' = \left(\frac{N_1}{N_2}\right)^2 \times R_2$$

In the circuit of the experiment

$$R' = \left(\frac{1}{2}\right)^2 \times 10 = 2\cdot5 \text{ ohms}$$

(b) Transformer stepping down the voltage between a supply and resistor.

Figure 6.20 shows the connections and circuit conditions. The 10 ohm resistor in the secondary circuit results in a resistance of 100 V/2·5 A = 40 ohms being presented to the supply.
It is clear that when a transformer which steps down voltage is inserted between the supply and the resistor, then the resistance presented to the supply is greater than that of the resistor.

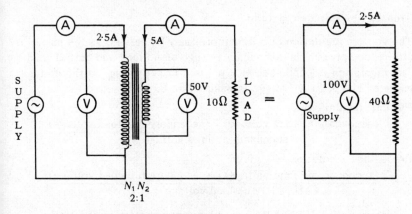

Fig. 6.20. Transferring resistance with a voltage step-down transformer.

If R'_2 is the equivalent resistance of the secondary winding R_1 when referred to the primary circuit we have

$$R'_2 = \left(\frac{N_1}{N_2}\right)^2 \times R_2$$

In this circuit

$$R'_2 = \left(\frac{2}{1}\right)^2 \times 10 = 40 \text{ ohms.}$$

If instead of dealing with the transfer of resistance from one side of the transformer to the other side we dealt with the transfer of reactance or impedance similar results would be forthcoming.

Let R'_2, X'_2 and Z'_2 be the equivalent values when referred to the primary circuit of R_2, X_2 and Z_2 in the secondary circuit
then

$$R'_2 = \left(\frac{N_1}{N_2}\right)^2 R_2$$

$$X'_2 = \left(\frac{N_1}{N_2}\right)^2 X_2$$

$$Z'_2 = \left(\frac{N_1}{N_2}\right)^2 Z_2$$

Thus we may now readily transfer resistance, reactance and impedance values from one side of a transformer to the other.

Transformer voltage regulation

The voltage regulation of a transformer may be defined as the numerical difference between the secondary voltage on no load and on full load. The regulation is expressed as a per unit or percentage of the no-load voltage. The primary voltage is assumed to be constant.

Per unit regulation =
$$\frac{\text{(secondary no-load voltage)} - \text{(secondary full-load voltage)}}{\text{secondary no-load voltage}}$$

Percentage reculation =
$$\frac{\text{(secondary no-load voltage)} - \text{(secondary full-load voltage)}}{\text{secondary no-load voltage}} \times 100.$$

With inductive loading the secondary voltage falls as the load increases and the regulation is positive, whereas with capacitive loads the secondary voltage may rise with increasing loads and the regulation is then negative.

Let us consider (Fig. 5.21) the simplified equivalent circuit of a transformer and the related phasor diagrams (Fig. 6.22). $\bar{R}_1$ represents both the resistance of the primary winding and the resistance of the secondary winding referred to the primary circuit. $\bar{X}_1$ and $\bar{Z}_1$ represent the total leakage reactance and impedance of the windings referred to the primary circuit.

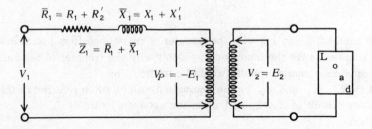

Fig. 6.21. Simplified equivalent circuit.

On no load the applied voltage V_1 is equal in magnitude to the e.m.f. induced in the primary winding. On load the applied voltage V_1 must overcome the back e.m.f. E_1 and also the voltage drops caused by the total winding resistances and reactances referred to the primary circuit of the transformer.

In Fig. 6.22(b) FA is drawn perpendicular to $-E_1$ extended. Triangles FAB and CDB are similar.

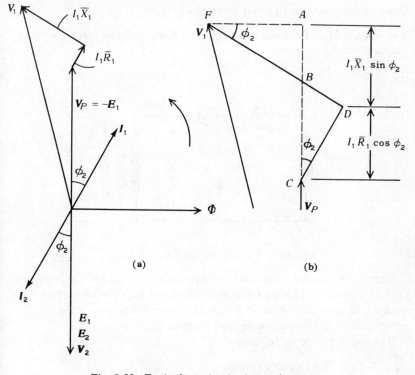

Fig. 6.22. Equivalent circuit phasor diagram.

$$V_1 - V_P \simeq AC$$
$$\simeq I_1 \bar{R}_1 \cos \phi_2 + I_1 \bar{X}_1 \sin \phi_2$$
$$\simeq I_1 (\bar{R}_1 \cos \phi_2 + \bar{X}_1 \sin \phi_2)$$

This expression gives a close approximation to the voltage drops produced by the winding impedance and hence enables us to calculate the transformer voltage regulation for loads of given power factor.

Instead of referring the winding impedances to the primary side of the transformer they may be referred to the secondary side, when the voltage regulation is given approximately by $I_2(\bar{R}_2 \cos \phi_2 + \bar{X}_2 \sin \phi_2)$

Let us now consider two simple tests commonly carried out on transformers to enable the voltage regulation and efficiency curves to be predicted.

Open circuit and short circuit transformer tests

Experiment 6 (v) *To predict the efficiency and voltage regulation of a single phase transformer*

(a) *Open circuit test*

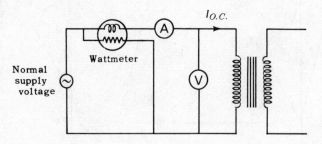

Fig. 6.23. Transformer open circuit test.

Normal voltage was applied to the primary winding and the resulting primary current and power input were observed with the secondary winding on open circuit as in Fig. 6.23. Corrections were made for power losses in the wattmeter (see Chapter 7).

The results were as follows:

Transformer rating;	2 kVA
Primary supply; ($V_{o.c.}$)	200 volts
Primary current; ($I_{o.c.}$)	0·71 ampere
Input power; ($P_{o.c.}$)	34·8 watts
Primary turns;	200
Secondary turns;	400

The input power on no load is assumed to supply the iron losses since with the small no-load current the $I_{o.c.}^2 R_1$ losses will be negligible.

(b) *Short-circuit test*

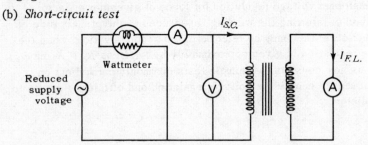

Fig. 6.24. Transformer short-circuit test.

A low voltage of normal frequency was applied and gradually increased until full load current $I_{F.L.}$ flowed in the short-circuited secondary winding (Fig. 6.24). Corrections were made for power losses in the instruments (see Chapter 7).
The results were as follows:

Primary supply; ($V_{S.C.}$)	17 volts
Primary current; ($I_{S.C.}$)	10 amperes
Secondary current; ($I_{F.L.}$)	5·2 amperes
Input power; ($P_{S.C.}$)	76 watts

In the short circuit test the losses in the iron core are negligible since with the low voltage applied to the primary winding the resulting flux density in the core will be small. It may be assumed therefore that the input power in this test supplies only the copper losses in the windings.

From the test results the total impedance, resistance and leakage reactance referred to the primary circuit of the transformer may be evaluated.

$$\bar{Z}_2 = \frac{V_{S.C.}}{I_{S.C.}} = \frac{17 \text{ V}}{10 \text{ A}} = 1\cdot 7 \text{ ohms}$$

$$\bar{R}_2 = \frac{P_{S.C.}}{I_{S.C.}^2} = \frac{76 \text{ W}}{(10 \text{ A})^2} = 0\cdot 76 \text{ ohm}$$

$$\therefore \bar{X}_1 = \sqrt{\bar{Z}_1^2 - \bar{R}_1^2}$$

$$\therefore \bar{X}_1 = \sqrt{1\cdot 7^2 - 0\cdot 76^2} = 1\cdot 52 \text{ ohms.}$$

$$\text{Efficiency} = \frac{\text{output}}{\text{input}} = \frac{\text{output}}{\text{output + losses}}$$

$$= \frac{V_2 I_2 \cos \phi_2}{V_2 I_2 \cos \phi_2 + \text{copper losses + iron losses}}$$

The iron losses as we have already remarked are constant but the copper losses depend on the (current)^2.

The following table which has been prepared from the test results enables us to plot a curve of efficiency against secondary load current when the power factor is 0·8.

Output current (% of full load)	A $V_2 I_2 \cos \phi_2$ (VA × 0·8)	B Iron loss (W)	C Copper loss (W)	Input A + B + C (W)	Efficiency $\frac{A}{A + B + C}$
120	1920	35	110	2065	93
100	1600	35	76	1711	93·5
90	1440	35	62	1537	93·8
80	1280	35	49	1364	93·8
70	1120	35	37	1192	94
60	960	35	27	1022	94
50	800	35	19	853	93·8
40	640	35	12	687	93
30	480	35	7	522	92
20	320	35	3	358	89·4
10	160	35	0·8	196	81·6

Figure 6.25 includes the predicted efficiency curve.

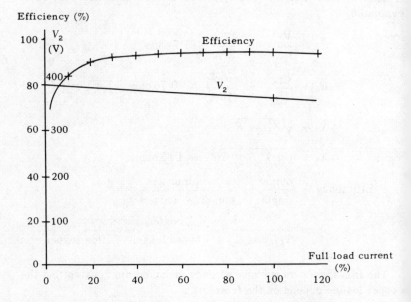

Fig. 6.25. Predicted voltage regulation and efficiency characteristics.

The voltage regulation is given approximately by,

$$I_1(\bar{R}_1 \cos \phi_2 + \bar{X}_1 \sin \phi_2)$$

From the short circuit test

$$\bar{R}_1 = 0.76 \text{ ohm}$$
$$\bar{X}_1 = 1.52 \text{ ohms}$$

Consider a secondary load with a power factor of 0·8

$$\begin{aligned}
\bar{R}_1 \cos \phi_2 + \bar{X}_1 \sin \phi_2 &= (0.76 \times 0.8) + (1.52 \times 0.6) \\
&= 0.61 + 0.91 \\
&= 1.52 \text{ ohms}
\end{aligned}$$

hence, full load voltage regulation = 10 A × 1·52 Ω = 15·2 volts.
Voltage applied to the primary winding

$$\begin{aligned}
&= 200 - 15.2 \\
&= 184.8 \text{ volts}
\end{aligned}$$

Voltage available at secondary terminals

$$\begin{aligned}
&= 2 \times 184.8 \\
&= 369.6 \text{ volts.}
\end{aligned}$$

As an alternative to using the primary values of equivalent resistance and reactance we could have worked in secondary values. Remembering that a winding which steps up voltage also steps up resistance and reactance we have,

$$\bar{R}_2 = \bar{R}_1 \left(\frac{N_2}{N_1}\right)^2 = 0.76 \times \left(\frac{400}{200}\right)^2 = 0.76 \times 4 = 3.04 \text{ ohms}$$

$$\bar{X}_2 = \bar{X}_1 \left(\frac{N_2}{N_1}\right)^2 = 1.52 \times \left(\frac{400}{200}\right)^2 = 1.52 \times 4 = 6.08 \text{ ohms}$$

$$\begin{aligned}
\bar{R}_2 \cos \phi_2 + \bar{X}_2 \sin \phi_2 &= (3.04 \times 0.8) + (6.08 \times 0.6) \\
&= 2.432 + 3.648 \\
&= 6.08 \text{ ohms.}
\end{aligned}$$

Therefore the full load voltage regulation = 5 A × 6·08 Ω = 30·4 V. The predicted curve for voltage regulation is also shown in Fig. 6.25.

Summary

Resistance reactance and impedance may be transferred from one side of a transformer to the other, and a transformer which steps up voltage also steps up resistance, reactance and impedance.

$$R'_2 = \left(\frac{N_1}{N_2}\right)^2 \times R_2$$

$$X'_2 = \left(\frac{N_1}{N_2}\right)^2 \times X_2$$

$$Z'_2 = \left(\frac{N_1}{N_2}\right)^2 \times Z_2$$

R_2 = secondary resistance
R'_2 = secondary resistance referred to the primary circuit.

The total resistance referred to the primary circuit,

$$\bar{R}_1 = R_1 + R'_2$$

Also $\quad \bar{X}_1 = X_1 + X'_2$

and $\quad \bar{Z}_1 = \bar{R}_1 + \bar{X}_1$

Voltage regulation $\simeq I_1(\bar{R}_1 \cos\phi_2 + \bar{X}_1 \sin\phi_2)$

The open circuit test enables us to determine the transformer core losses and the short circuit test enables us to calculate the total equivalent winding resistance, leakage reactance and impedance.

$$\text{Efficiency} = \frac{\text{output}}{\text{output} + \text{copper losses} + \text{iron losses}}$$

Examples 6.3.

1. Show how the equivalent circuit of a transformer may be reduced to an impedance consisting of resistance and inductive reactance.

2. Explain from first principles how resistance may be transferred from the secondary side of a transformer to the primary side.

A transformer has a secondary high-voltage winding with a resistance of 1·2 ohms. What would be the equivalent resistance in the primary circuit? The turns ratio is 1100/440.

3. A transformer has a primary winding resistance of 0·5 ohm and a leakage reactance of 0·4 ohm. The secondary winding with twice as many turns as the primary has a resistance of 0·4 ohm and a leakage reactance of 0·3 ohm. Calculate the total resistance, reactance and

impedance referred to the primary circuit.

4. Show by means of a phasor diagram the phase relationships between the input current and voltage, the magnetic core flux, and the output current and voltage of an ideal transformer operating with a resistive load.

An ideal transformer with a primary winding of 1250 turns has a 10 ohm load resistor across the secondary winding. A 100 V 50 Hz input produces 75 V across the load. Determine the number of secondary turns, the primary and secondary currents and the impedance presented to the supply.

5. Describe the short-circuit test as applied to a single-phase transformer. In an open-circuit test on a single-phase transformer the input power was observed to be 75 watts and the primary current 0·43 ampere when the transformer was connected across the normal rated supply of 240 volts. Determine the values of two components in the equivalent circuit.

6. How does the short-circuit test of a transformer enable us to determine the equivalent values of resistance reactance and impedance of the transformer windings?

With the secondary winding of a transformer short-circuited and a reduced voltage of 9 volts applied to the primary winding it was observed that a full load secondary current of 20 amperes resulted. The primary current was 10·3 amperes and the input power 72 watts. Calculate the values of appropriate components in the equivalent circuit.

7. What is meant by the percentage regulation of a transformer?
A 10 kVA 50 Hz 440/240 V single-phase transformer takes 140 W at a current of 0·3 A when 240 V is applied to the low voltage winding and the high voltage side is on open circuit. With the low voltage winding short-circuited and 31 volts applied across the high-voltage winding the current in this winding is 25 A and the power input 200 W. Calculate the efficiency and voltage regulation when the transformer is on full-load at a power factor of 0·5 lagging.

8. Sketch a complete phasor diagram for a single-phase transformer supplying a load with lagging current and explain the causes of the voltage changes with varying loads.

A 75 kVA 1100/400 V single-phase transformer takes 1000 W on no load. Full load current is produced in the windings when 23 V is applied to the high-voltage winding and the low-voltage side is short-circuited. The corresponding power loss is 1050 W. Calculate the efficiencies at both full load and half load with unity power factor and the percentage full-load voltage regulation at a lagging power factor of 0·8.

Separating transformer core losses

The losses in a transformer core may be divided into eddy current and hysteresis loss components.

(a) *Eddy current losses*

Whenever the flux linking with a closed circuit changes, an e.m.f. is induced.

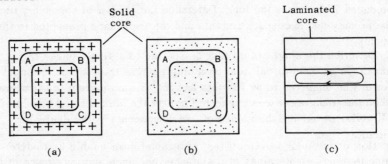

Fig. 6.26. Sections through solid and laminated transformer cores.

Consider a section through the iron core of a transformer limb which carries an alternating flux (Fig. 6.26(a & b)). In any closed circuit such as $ABCD$ the alternating flux linking with this circuit induces an e.m.f. which produces an alternating current termed an eddy current. As the eddy currents circulate they produce heat in the transformer core.

The induced e.m.f. in any path is directly proportional to the magnitude of the flux density (B) and frequency (f) of the flux wave. Since the eddy current power loss (P_E) in a circuit of resistance R is given by

$$\frac{(\text{induced e.m.f.})^2}{R}$$

the eddy current loss $\propto B^2$ and f^2

$$P_E \propto B^2 f^2$$

To limit the eddy current power losses to a reasonable value transformer cores are laminated (Fig. 6.26(c)).

The laminations are insulated from each other and this reduces individual induced e.m.f.'s and circulating currents.

(b) *Hysteresis core loss*

The hysteresis loss in iron subject to alternating magnetisation, has already been dealt with (Volume 1, Chapter 6). For a given maximum flux density it was seen that the hysteresis power loss,

$$P_H \propto f$$

The experiment which follows shows how the iron losses in a transformer core may be separated into these component losses.

Experiment 6(vi) *To seperate the eddy current and hysteresis losses in a transformer core*

The e.m.f. equation for the primary winding of a transformer may be written,

$$V_1 \simeq E_1 = 4 \cdot 44 f \Phi_m N_1 \text{ volts}$$

and since $B_m = \Phi_m/a$ we see that the maximum flux density in a transformer core depends on the applied voltage.

For a constant applied voltage magnetic hysteresis results in a power loss $P_H \propto f$. hence $P_H = kf$ where k is a constant.

It has already been seen that the eddy current power loss depends on the (core circulating e.m.f.'s)^2. These e.m.f.'s are proportional to the frequency, and for a constant core flux density the eddy current power loss, $P_E = k'f^2$.

Hence the total iron losses, $P = kf + k'f^2$ providing B_m is maintained constant.

Now, $P/f = k + k'f$ is the equation of a straight line. By varying the frequency of the voltage applied to a transformer on no load and measuring the input power which is assumed to supply the iron losses, we may draw the line represented by this equation and hence determine the constants k and k'. It is essential however that the core flux density is maintained constant.

Since

$$V_1 = 4 \cdot 44 f \Phi_m N \text{ volts}, \quad \Phi_m \propto \frac{V_1}{f}$$

Thus if the ratio V_1/f is maintained constant the maximum flux and hence the flux density must remain constant.

The primary winding of the single phase transformer under test was connected across the output terminals of one phase of a three-phase alternator (Fig. 6.27) driven by a variable speed motor.

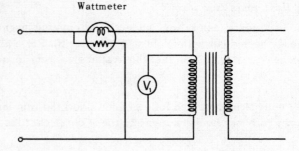

Variable frequency supply

Fig. 6.27. Separating hysteresis and eddy current iron losses.

The transformer was rated for use on a 250 V 50 Hz supply and the ratio V_1/f was maintained at 250 V/50 Hz throughout the test. The results were as tabulated.

V_1 (V)	$\dfrac{V_1}{f}$	f (Hz)	P (W)	$\dfrac{P}{f}$
278	5	55·5	56	1·01
266·5	5	53·4	52·9	0·99
258	5	51·6	50·5	0·975
250	5	50·0	48·6	0·972
244	5	48·9	47·2	0·965
230	5	46	43·2	0·94

From the graph of P_f plotted against frequency, k and k' were evaluated (Fig. 6.28).

Since $P = kf + k'f^2$ we have $P = 0·6f + 0·0074f^2$.

By taking increasing values for f we may tabulate the hysteresis and eddy current losses over a selected frequency range.

f (Hz)	P_H (W)	P_E (W)	P (W)
10	6	0·74	6·74
20	12	2·86	14·86
30	18	6·66	24·66
40	24	11·9	35·9
50	30	18	48
60	36	26·6	62·6

The curves in Fig. 6.29 show how the core losses vary with increasing frequencies.

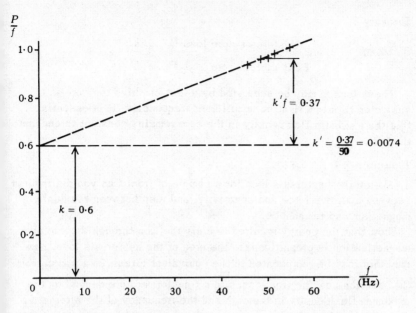

Fig. 6.28. Evaluating the hysteresis and eddy current constants.

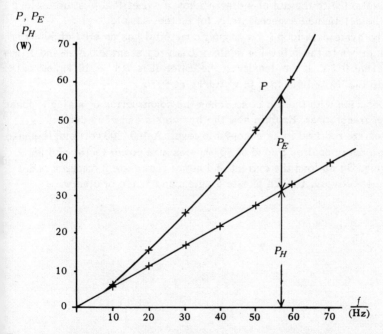

Fig. 6.29. Variation of hysteresis and eddy current losses with frequency.

233

Summary

$$\text{Hysteresis core loss } P_H = kf$$
$$\text{Eddy current core loss } P_E = k'f^2$$

These losses may be separated by a test in which the core is subjected to magnetisation at different frequencies. It is essential that the maximum flux density in the core remains constant throughout the test.

Examples 6.4.

1. Sketch the hysteresis loop for a sample of iron. Can you distinguish between coercive force and coercivity, and also between residual magnetism and remanence?

Show that the energy required to carry the iron through a cycle of magnetisation is proportional to the area of the hysteresis loop. How may this loss be represented in the equivalent circuit of a transformer?

2. Describe how the iron losses in a transformer core depend on the maximum flux density in the core and the frequency of the alternating flux. Explain how the iron losses in the core may be separated.

3. Explain with the aid of a diagram how a symetrical hysteresis loop may be determined experimentally for a steel sample.

The hysteresis loop for a magnetic material has an area of 5000 mm² when drawn to the following scales: 10 mm represent 200 A/m and 10 mm represent 0·1 T. If the density of the material is $7 \cdot 8 \times 10^2$ kg/m³ calculate the hysteresis loss in watts/kg at 50 Hz.

4. Describe with the aid of sketches the construction of a single-phase power transformer. Explain how the hysteresis and eddy-current losses are reduced to an acceptable level. A 400/200 volt single-phase transformer supplies a load of 50 amperes at a power factor of 0·85 lagging. On no load the current and power factor are 3 amperes and 0·2 respectively. Draw a phasor diagram and hence or otherwise

determine the current taken from the supply and the power factor on load. Neglect winding resistances and leakage reactances.

5. Enumerate and explain the losses which occur in a single-phase transformer on load.

A single-phase 200 kVA 200 V transformer has a full-load efficiency of 96% at unity power factor when the ratio of the copper losses to the iron losses is 3 : 1. Determine the full load iron and copper losses.

Auto transformers

The transformers used in previous tests had separate primary and secondary windings. We now come to consider auto transformers with only one winding which serves both as the primary and as the secondary windings.

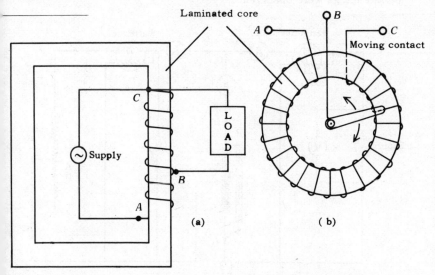

Fig. 6.30. Auto transformer connections.

An auto transformer with a single winding on a laminated iron core is shown in Fig. 6.30 (a). The whole of this winding normally serves as the transformer primary winding to produce an alternating flux in the laminated core.

Part of the winding, such as the section $B - C$, serves also as a secondary winding.

Figure 6.30(b) represents a continuously variable-ratio auto transformer. A movable tapping point makes contact with the turns of a toroidal coil wound on a laminated iron core.

Experiment 6(vii) *To investigate the currents in the various sections of an auto transformer winding when supplying a load current*

The auto transformer (Fig. 6.31) used in this test had a cylindrical laminated core carrying a single layer toroidal winding. The part of the winding between the movable contact C and end B served also as the secondary winding of the transformer. When the rated voltage of 240 V was applied to the primary winding the nominal voltage appearing across B and C for any position of the tapping point C was engraved on a circular scale.

In the test the position of C was adjusted to give a nominal output of 100 volts on no load with the transformer connected across a 240 volt supply. The secondary load current was varied in steps of about 1 ampere by adjusting the secondary circuit load resistor. All instrument readings were observed.

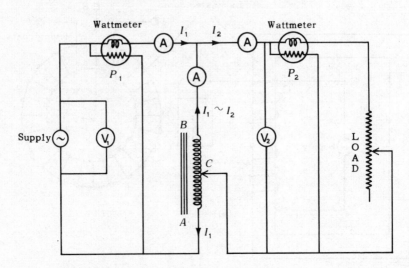

Fig. 6.31. Auto transformer load test circuit.

The instrument readings were as tabulated.

V_1 (V)	I_2 (A)	I_1 (A)	$I_1 \sim I_2$ (A)	V_2 (V)	P_1 (W)	P_2 (W)	$\cos\phi_1$ $P_1/V_1 I_1$	$\cos\phi_2$ $P_2/V_2 I_2$	η (%)
240	—	0.32	0.32	100	30	—	—	—	—
240	1.05	0.65	0.6	99.5	150	108	0.96	1.0	72
240	2.0	1.02	1.08	99	236	206	0.96	1.0	87.4
240	3.0	1.44	1.62	98	332	294	0.96	1.0	88.8
240	4.0	1.81	2.2	97.5	430	390	0.99	1.0	90.8
240	5.0	2.25	2.78	97	530	488	0.99	1.0	92.2
240	6.0	2.7	3.35	97	636	586	0.98	1.0	92.2
240	7.0	3.12	3.98	96.5	740	690	0.98	1.0	93.2
240	8.0	3.57	4.5	96	840	780	0.98	1.0	93.0

The curves in Fig. 6.32 show how the efficiency, secondary voltage, and the current in the winding common to the primary and secondary circuits vary as the load increases.

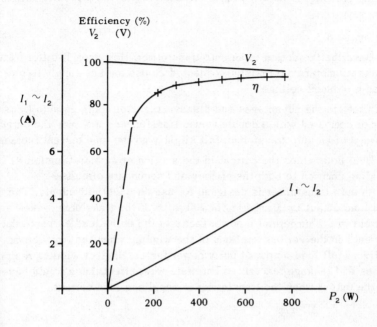

Fig. 6.32. Auto transformer characteristics.

The supply current to an auto transformer includes the no-load current. The supply current must also establish the ampere turns in the primary winding, to balance the ampere turns resulting from the load current in that part of the winding which acts as the secondary winding of the transformer.

The primary and secondary currents I_1 and I_2 in the section of the winding common to both circuits, are almost in phase opposition. The resultant current in this section is the phasor combination $I_1 + I_2$.

The main advantage of the auto transformer is that it requires less copper than a double wound transformer. The closer the voltage ratio of the primary and secondary circuits is to unity then the greater is this saving. Auto transformers have a disadvantage that because of the common connection between the primary and secondary circuits these circuits cannot be isolated.

Auto transformers have many applications in voltage regulating devices where small degrees of bucking or boosting of voltage are required. They may be used to provide reduced voltages to a.c. motors during starting. Laboratory units are used to provide smoothly variable a.c. supplies.

Examples 6.5.

1. Describe the action of an auto transformer. How does it differ from a potentiometer type resistor connected across an a.c. supply to produce a reduced voltage?

2. What are the advantages and disadvantages of using an auto transformer compared with a double wound transformer? Why may the output voltage of an auto transformer fall slightly as the load current increase

3. What determines the current in the section of an auto transformer winding common to both the primary and secondary circuits?

An auto transformer is designed for use on a 240 volt supply. The full load output is 20 amperes at 200 volts. If the transformer takes a current of 0·8 ampere at a power factor of 0·3 on no load estimate the currents in the various sections of the winding when the transformer delivers full load output at unity power factor. Neglect winding resistances and leakage reactances but state what effect these would have on the output when the transformer is supplying load current.

7. The Simple Potentiometer and Electrical Measuring Instruments

Introduction

You have now been introduced to many aspects of electrical technology and have studied experiments illustrating the principles and applications of fundamental theory. During earlier studies use has been made of indicating instruments including ammeters, voltmeters and wattmeters. Later in this chapter we are going to examine some of the instruments used in electrical measurements work but first we will consider the simple potentiometer with its particular application to the checking of instrument calibrations.

Connecting and balancing a simple potentiometer

Experiment 7(i) *To connect a simple potentiometer and to determine the condition for balance*

In this experiment a circuit was connected as shown in Fig. 7.1. A 2 volt lead-acid cell (B) was connected across a length of resistance wire $M - N$. The wire had a uniform cross-sectional area and was stretched alongside a scale.

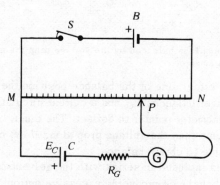

Fig. 7.1. Simple potentiometer.

A single dry cell (C) from a torch battery and a centre-zero galvanometer with a protective series resistor were connected to the terminal M. The adjustable slider P could be moved along the resistance wire to make contact at any point. It is important to note that the positive terminals of B and C were both connected to M.

The switch S was closed and it was observed that when contact was momentarily made between P and the slide wire, the galvanometer pointer was deflected. As P was moved along the wire a point was reached where the galvanometer deflection was zero. At either side of this point the galvanometer deflections were in opposite directions. The point where the deflection is zero is termed the balance point.

Now, why is the galvanometer deflection zero at one particular point?

Current flows from the cell B along $M - N$ and consequently there is a voltage drop V_{MP} from M to P (Fig. 7.2). If this voltage drop is exactly equal to the e.m.f. E_C of the cell C then there will be no resultant e.m.f. acting in the circuit MPO, no current will flow through the galvanometer, and the galvanometer deflection will be zero.

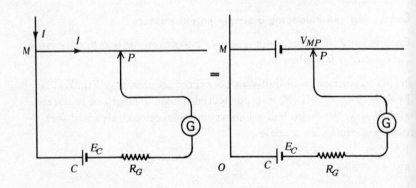

Fig. 7.2. When the e.m.f. is balanced by the voltage drop the galvanometer deflection is zero.

As P is moved either side of the balance position the voltage drop along $M-P$ no longer balances E_C and a current circulating in MPO causes the galvanometer pointer to deflect. The e.m.f. of the cell C must always be less than the voltage drop along $M-N$, otherwise it will not be possible to obtain balance.

A resistor R_G is included in series with the galvanometer to protect it in the event of unbalance producing excessive currents.

It will be seen in the next experiment that we must have a reference standard of e.m.f. with which to calibrate the potentiometer. First, however we will consider one type of standard cell.

Standard cell

The Weston standard cell is commonly used as a reference standard when measurements are made with a potentiometer.

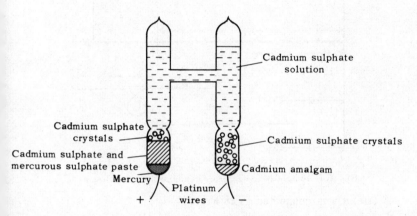

Fig. 7.3. Weston standard cell.

The positive electrode is mercury and this is covered with a paste of cadmium and mercurous sulphate which acts as a depolarising agent (see Chapter 11). The negative electrode is an amalgam of cadmium with mercury. A solution of cadmium sulphate constitutes the electrolyte and this solution is maintained saturated by crystals of cadmium sulphate held in both limbs by constrictions in the glass tubes of the cell.

The e.m.f. of the cell depends on the purity of the constituent materials, but when the cell is manufactured according to a prescribed specification it has an e.m.f. of 1·01859 volts. Changes in temperature have little effect on the e.m.f. but if necessary, corrections may be applied. The cell has a high internal resistance of 900–1000 ohms and is only suitable for use as a standard of e.m.f.

Comparing the e.m.f.'s of two cells

Now that you appreciate the meaning of balance in a potentiometer circuit and understand the construction of a standard cell, let us see how the potentiometer may be applied to determine the unknown e.m.f. of a cell.

Experiment 7(ii) *To connect a simple potentiometer to measure the e.m.f. of a cell*

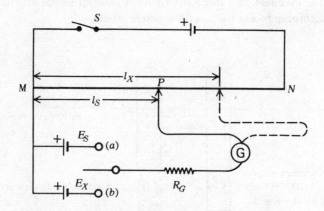

Fig. 7.4. Simple potentiometer connected to measure e.m.f.

A circuit was connected as in Fig. 7.4.

In this circuit E_S was a Weston standard cell and E_X was the cell of unknown e.m.f.

With the switch S closed and the two-way switch in position (a) the slider P was adjusted until the galvanometer deflection was zero. The distance $M-P$ was l_S metre.

If ρ is the resistivity of the slide wire material and a is the cross-sectional area of the slide wire then with a current I,

$$V_{MP} = I \times \frac{\rho \times l_S}{a}$$

and this voltage drop is balanced by the e.m.f. E_S.

The two-way switch was then moved to position (b) and the balance point determined with the unknown cell connected across $M-P$. Balance was obtained with P at a distance of l_X metre from M.

Thus

$$E_X = I \times \frac{\rho l_X}{a}$$

Now, since the wire is of uniform cross-sectional area we may write

$$\frac{E_S}{E_X} = \frac{l_S}{l_X} \quad \text{or} \quad E_X = E_S \times \frac{l_X}{l_S} \text{ volts.}$$

All quantities except E_X are known and therefore the value of the unknown e.m.f. may be calculated.

It is advisable after determining l_X to return the two-way switch to position (a) and check that balance is still obtained when the standard cell is connected in the circuit with M–P equal to l_S metre.

In determining the distance l_S we are standardising the potentiometer, or determining the voltage drop per unit length of M–N, as the current flows through the wire.

The observations made were as follows:
Standard cell: Weston Cadmium,

$$E_S = 1\cdot 01859 \text{ volts}$$
$$l_S = 0\cdot 504 \text{ metre}$$

Unknown cell: Leclanché,

$$l_X = 0\cdot 705 \text{ metre}$$

$$E_X = E_S \times \frac{l_X}{l_S} = 1\cdot 01859 \times \frac{0\cdot 705}{0\cdot 504} \text{ volts}$$

$$\therefore E_X = 1\cdot 425 \text{ volts}$$

The voltage drop along the wire is

$$\frac{1\cdot 01859 \text{ V}}{0\cdot 504 \text{ m}} = 2\cdot 02 \text{ V/m}$$

Notes

(1) The total voltage drop along M–N, which is one metre in length, is 2·02 volts and providing that the e.m.f. being measured is less than 2·02 volts it will be possible to adjust the slider P until balance is obtained. Higher e.m.f.'s are said to be outside the slide wire range and cannot be measured directly.

(2) You may ask 'Why go to all the trouble of using a potentiometer to measure the e.m.f. of a cell when a voltmeter could be used?' The answer is that a voltmeter does not normally measure e.m.f. A voltmeter measures the p.d. across the terminals of a cell (Fig. 7.5) and if the voltmeter requires a current — however small this may be — to operate it then the voltmeter indicates,

$$\text{p.d.} = \text{e.m.f.} - (I \times r)$$

where r is the internal resistance of the cell and I the current taken by the voltmeter.

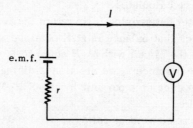

Fig. 7.5. A voltmeter indicates p.d. and not e.m.f.

When a potentiometer is balanced there is no current through the cell under test and hence there can be no internal voltage drop of $I \times r$ volts. The potentiometer measures true e.m.f.

One method of determining the internal resistance of a cell or source of e.m.f. is to measure the e.m.f. of the source using a potentiometer and then measure the potential difference across the terminals of the cell when a resistor of known value is connected across the cell terminals. Here is an example.

The e.m.f. of a cell measured by means of a potentiometer was 2 volts. A 12 ohm resistor was then connected across the terminals of the cell and the potential difference across the cell when measured with a potentiometer was 1·83 volts.

Determine the internal resistance of the cell.

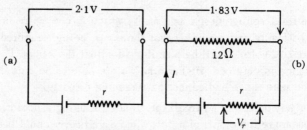

Fig. 7.6. Example to determine internal resistance.

In diagram (a)   e.m.f. = 2·1 V
In diagram (b)   p.d. = 1·83 V

$$I = \frac{1 \cdot 83 \text{ V}}{12 \, \Omega} = 0 \cdot 1525 \text{ A}$$

$$V_r = 2 \cdot 1 \text{ V} - 1 \cdot 83 \text{ V} = 0 \cdot 27 \text{ V}$$

Internal resistance,

$$r = \frac{V_r}{I} = \frac{0 \cdot 27 \text{ V}}{0 \cdot 1525 \text{ A}}$$

$$\therefore r = 1 \cdot 77 \text{ ohm}$$

Measurement of current by means of a potentiometer

We have seen how the simple potentiometer may be used to measure e.m.f. Let us now look at an experiment where it is used to determine the current flowing in a circuit.

Experiment 7 (iii) *To use a simple potentiometer to determine the current flowing in a d.c. circuit*

A circuit was connected which enabled us to determine the current flowing through a lamp when the p.d. across the terminals was 6 volts (Fig. 7.7).

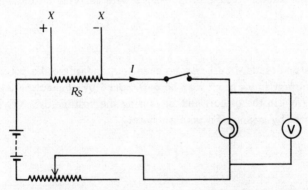

Fig. 7.7. Determining current by measuring the voltage drop across a standard resistor.

A standard resistor R_S of 0·1 ohm was connected in series with the lamp and a variable resistor adjusted until the p.d. across the lamp terminals was 6 volts.

The current through the lamp was determined by measuring the voltage drop V_R across the terminals $X-X$ of the standard resistor.

$$I = \frac{V_R}{R_S}$$

Students are advised to sketch a complete circuit including the potentiometer shown in Fig. 7.4, but with the cell of unknown e.m.f. replaced by the voltage drop across the terminals $X-X$ (Fig. 7.7).

In a particular test carried out the following observations were made:
length

l_S = 0·51 metre to balance E_S of 1·01859 volts

l_X = 0·202 metre to balance the p.d. across $X-X$

R_S = 0·1 ohm

Thus the p.d. across $X-X = \dfrac{1 \cdot 01859}{0 \cdot 51} \times 0 \cdot 202 = 0 \cdot 403$ volt.

Hence the current

$$I = \frac{0 \cdot 403 \text{ V}}{0 \cdot 1 \, \Omega} = 4 \cdot 03 \text{ ampere}$$

Summary

When using a simple potentiometer the e.m.f. to be measured is balanced against the known voltage drop along a wire carrying current.

$$\frac{E_S}{E_X} = \frac{l_S}{l_X}$$

A Weston standard cell may be used to standardise the potentiometer.

The current in a circuit may be determined by connecting a standard resistor (R_S) in the circuit and measuring the voltage drop (V_R) across the resistor by means of a potentiometer.

$$I = \frac{V_R}{R_S}$$

Examples 7.1.

1. Describe how the simple potentiometer may be used to compare the e.m.f.'s of two cells.

In an experiment using a simple potentiometer balance was obtained when a standard cell of e.m.f. 1·018 volts was connected across 0·368 m of the potentiometer wire. Calculate the e.m.f. of a second cell if balance is obtained when it is connected across 0·552 m of the wire.

What would be the disadvantage of using a voltmeter in the above experiment compared with a potentiometer?

2. A potentiometer wire A-B is 1·0 m long, has a resistance of 6·2 ohms, and is connected to a battery through an ammeter and a control resistor. The current is adjusted to 0·5 ampere. A second circuit having a total resistance of 80 ohms includes a galvanometer in series with a cell of e.m.f. 2·01 V. This circuit is connected between the end A of the potentiometer and a sliding contact C. Determine the position of the sliding contact for zero galvanometer deflection. Calculate the current through the galvanometer if the sliding contact C is moved 20 mm from the balance position towards the end B. Assume that the battery current remains unchanged. Draw a circuit diagram showing the currents in each part of the circuit.

3. With the aid of a circuit diagram show how the internal resistance of a cell may be determined using a simple slide-wire potentiometer.

A 2 metre potentiometer incorporating a standard cell of e.m.f. 1·0183 V was balanced with the slider adjusted to a position 1·0 m from the positive end of the wire. With a cell of unknown e.m.f. replacing the standard cell balance was obtained at 1·53 m from the same end and when the cell was delivering a load current of 1 A the balance point was 1·43 m. Calculate the internal resistance of the cell.

4. Describe the construction of a standard cell for use with a potentiometer.

A cell on open circuit balanced across 0·76 m of wire in a certain potentiometer and across 0·608 m when a 4 ohm resistor was connected across the cell terminals. What is the internal resistance of the cell?

5. How may a simple potentiometer be connected to enable us to determine the current in a circuit?

A potentiometer consists of a one metre length of wire and when a standard cell of e.m.f. 1·018 V is connected, balance is obtained with the tapping point on the wire 0·274 m from the end to which the standard cell is connected. If the voltage drop across a 0·1 ohm resistor carrying an unknown current is balanced when the tapping point on the wire is 0·765 m from the same end of the wire calculate the value of the unknown current.

6. Describe with the aid of a diagram how you would (a) calibrate a simple slide wire potentiometer and (b) use a simple potentiometer to determine the value of a resistor of the order of 1 ohm.

7. Explain the principle of the simple potentiometer when used as a means of comparing e.m.f.'s. Why is it important that the slide wire should be uniform throughout its length?

A small direct current flows through a 15 ohm resistor. How could a potentiometer be used to measure this current without breaking the circuit?

If the standard cell in the potentiometer circuit has an e.m.f. of 1·018 V and gives a balance against half the length of the slide wire what proportion of the length would correspond to the measurement of 22 mA in the 15 ohm resistor?

8. Can you suggest how the accuracy of an ammeter could be checked using a standard resistor of 1 ohm and a simple potentiometer? Give a circuit diagram.

In a particular test, balance was obtained with a standard cell of 1·018 volts connected across 0·457 m of the slide wire. When the ammeter current was adjusted to 0·5 ampere, the voltage drop across the 1 ohm standard resistor was balanced with the tapping point on the slide wire at a corresponding distance of 0·221 m. Calculate the error in the ammeter reading.

9. A circuit (Fig. 7.8) is used to measure small changes in a current I. G is a sensitive galvanometer and E is a battery with a constant e.m.f. of 2·02 volts.

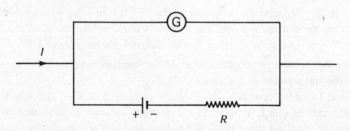

Fig. 7.8. Example 9.

Determine the value of the resistor R necessary to reduce the galvanometer current to zero when the current I is 0·2 ampere.

A commercial form of the direct current potentiometer used to check the calibration of a voltmeter

The simple potentiometer used in the previous experiments has disadvantages in that arithmetical calculations are involved in evaluating the results. The accuracy with which the balance point can be obtained is limited by the length of the slide wire that can be mounted, and it is essential that the slide wire is of uniform cross-section

throughout its length. These disadvantages have led to the design of
a d.c. potentiometer which is practically direct reading and the general
arrangement of this form of potentiometer is dealt with in the experiment
which follows.

Experiment 7(iv) *To check the calibration of a d.c. voltmeter using a direct reading potentiometer*

A simplified diagram of the potentiometer circuit is shown in Fig. 7.9.

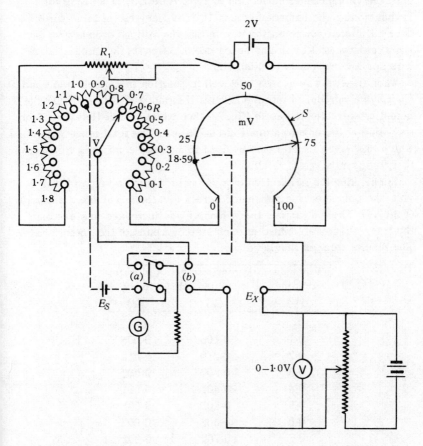

Fig. 7.9. Direct reading potentiometer connected to check the calibration of a 0–1·0V voltmeter.

The slide-wire (S) is in series with eighteen resistors (R) each having resistance equal to that of the slide wire. These additional resistors have the effect of artificially increasing the length of the slide wire.

Current flows through the slide wire from a 2 volt source and may be varied by a variable resistor R_1. A scale marked with 100 divisions is adjacent to the slide wire.

When the two-way switch is set in position (*a*) a standard cell (E_S) of e.m.f. 1·01859 volts is connected to fixed points as shown by the broken line. By varying R_1 the current through the slide wire is adjusted until the galvanometer deflection is zero. An e.m.f. of 1·01859 volts is then seen to be balanced across 10·1859 'lengths' of slide wire. We have calibrated our potentiometer so that the voltage drop across each fixed resistor is 0·1 volt and hence one division on the circular slide wire scale corresponds to 0·001 volt.

When the two way switch is moved to position (*b*) an unknown e.m.f. E_X may be introduced into the circuit. If balance is obtained with the potentiometer adjusted as in Fig. 7.9 then we may immediately read off the value of the unknown e.m.f. as 0·775 volt. The total range within which this potentiometer may be used to measure e.m.f.'s is from zero to 1·9 volts in steps of 0·001 volt.

In checking the calibration of a voltmeter having a range of 0–1·0 volt, the voltmeter was connected across a variable voltage d.c. supply (Fig. 7.9). The p.d. across the voltmeter was increased in steps and the true voltage determined at each step by means of the potentiometer. The results were as tabulated.

Voltmeter reading (V)	Potentiometer measurement (V)	Voltmeter error (V)
0	–	–
0·1	0·106	0·006
0·2	0·210	0·010
0·3	0·302	0·002
0·4	0·401	0·001
0·5	0·501	0·001
0·6	0·602	0·002
0·7	0·702	0·002
0·8	0·801	0·001
0·9	0·905	0·005
1·0	1·004	0·004

It will be observed that the voltmeter indicated values throughout the test slightly below the true values. Fig. 7.10 shows how the errors vary. An error is regarded as positive if the instrument reads in excess of the true value, and negative if it indicates less than the true value.

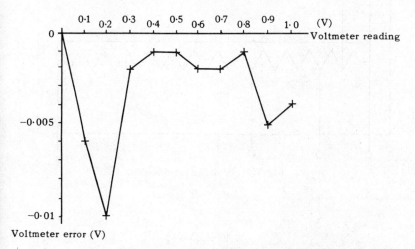

Fig. 7.10. Voltmeter calibration curve.

The general principle upon which the d.c. potentiometer operates has been dealt with but it must be pointed out that commercial instruments include additional refinements. In particular, additional resistor networks are provided which may be connected across the slide wire circuit to reduce the slide wire current and thus enable the potentiometer to be used to measure voltages in lower ranges. However, we are now going to study the extension of the upper limit of the potentiometer range.

Extending the range of a potentiometer

It is clear from our preceding work that the highest e.m.f. which may be measured by a potentiometer is limited by the voltage drop along the potentiometer wire. To measure voltages which are in excess of this value we must devise a method of taking and measuring a known portion of the unknown voltage.

A voltage ratio box as shown in Fig. 7.11 may be used.

The voltage V which is to be measured is connected across a high resistor of R ohms and a known portion of V is applied to the potentiometer.

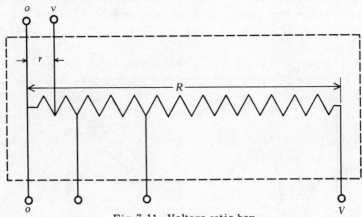

Fig. 7.11. Voltage ratio box.

$$\frac{V}{\nu} = \frac{R}{r}$$

$$\therefore V = \frac{R}{r} \times \nu$$

By providing various tappings on the volt-box a number of different ratios may be obtained.

In a particular unit examined R had a value of 20 000 ohms and tappings were provided which enabled ratios of 200, 100, 80, 40, 20, 10, 8, 4 and 2 to 1 to be obtained.

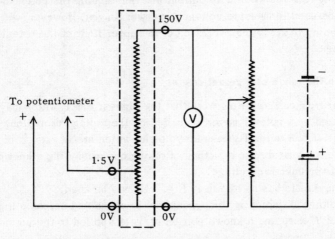

Fig. 7.12. Checking the calibration of a voltmeter.

Experiment 7 (v) *To connect a direct reading potentiometer to check the calibration of a 0–150 V voltmeter*

The voltmeter under test was connected across a d.c. supply as shown in Fig. 7.12.

A volt-box with a ratio of 150/1·5 was connected between the voltmeter and the E_X terminals of a direct reading potentiometer (Fig. 7.12). The output from the volt-box was within the range of the potentiometer and by measuring this output carefully the true p.d. across the voltmeter could be determined.

The p.d. across the voltmeter terminals was increased in steps, the output from the volt-box measured, and errors in the instrument reading calculated.

Volt-box ratio 150/1·5

Voltmeter reading (V)	Potentiometer measurement (V)	True p.d. (V)	Voltmeter error (V)
0	0	0	0
10	0·101	10·1	−0·1
20	0·201	20·1	−0·1
30	0·301	30·1	−0·1
40	0·409	40·9	−0·9
50	0·505	50·5	−0·5
60	0·606	60·6	−0·6
70	0·704	70·4	−0·4
80	0·802	80·2	−0·2
90	0·903	90·3	−0·3
100	1·006	100·6	−0·6
110	1·091	109·1	+0·9
120	1·193	119·3	+0·7
130	1·292	129·2	+0·8
140	1·393	139·3	+0·7
150	1·485	148·5	+1·5

The error is shown plotted against instrument reading in Fig. 7.13.

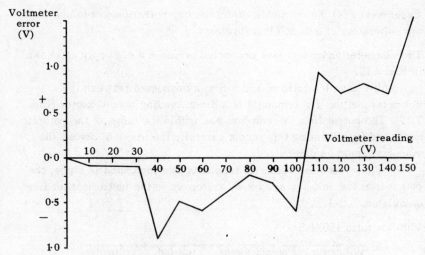

Fig. 7.13. 0–150 volt voltmeter calibration curve.

Checking the calibration of an ammeter

Having checked the calibration of voltmeters within and outside the slide wire range we now come to the checking of an ammeter calibration.

Experiment 7 (vi) *To check the calibration of an ammeter using a direct reading potentiometer*

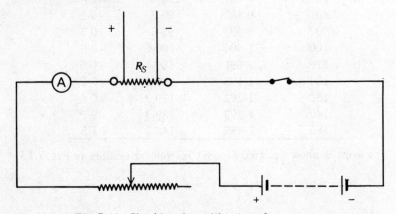

Fig. 7.14. Checking the calibration of an ammeter.

The ammeter under test was connected in series with a standard resistor R_S, a variable resistor and a d.c. supply as in Fig. 7.14. The true current was determined by measuring the voltage drop across R_S.

When the ammeter carries a current which produces full scale deflection of the pointer, the voltage drop across the potential terminals of the standard resistor must be within the range of the potentiometer.

In this experiment a 0–5 A ammeter was checked and the standard resistor chosen was of 0·1 ohm. This gives an expected maximum voltage drop to be measured by the potentiometer of about 0·5 V when the ammeter indicates 5 amperes.

The observations made as the current through the ammeter was increased in steps were as tabulated.

Ammeter reading (A)	Potentiometer measurement (V)	True current (A)	Ammeter error (A)
0·0	0·0	0·0	0·0
0·5	0·046	0·46	+0·04
1·0	0·097	0·97	+0·03
1·5	0·149	1·49	+0·01
2·0	0·199	1·99	+0·01
2·5	0·249	2·49	+0·01
3·0	0·298	2·98	+0·02
3·5	0·349	3·49	+0·01
4·0	0·398	3·98	+0·02
4·5	0·448	4·48	+0·02
5·0	0·501	5·01	−0·01

The error curve is shown plotted against instrument reading in Fig. 7.15.

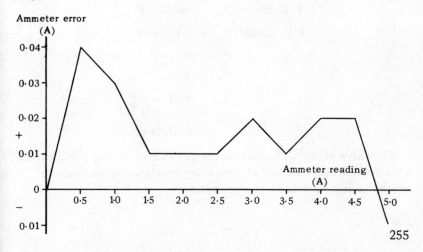

Fig. 7.15. 0–5 A ammeter calibration curve.

Measuring low resistance with the aid of a potentiometer

Here is a further application of the potentiometer.

Experiment 7(vii) *To measure low resistance using a potentiometer*

In this experiment the resistance between the potential terminals of an ammeter shunt was to be measured. A circuit was connected as shown in Fig. 7.16.

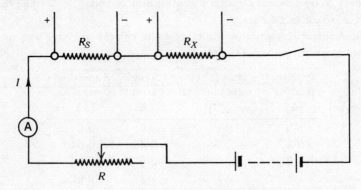

Fig. 7.16. Determining resistance by comparing voltage drops.

The unknown resistor R_X was connected in series with a standard resistor R_S and by adjusting a variable resistor R the current I was set to a suitable value. The voltage drops across R_S and R_X were then measured using a standardised d.c. potentiometer.

Observations were as follows:

$$R_S = 0.01 \text{ ohm}$$
$$\text{P.D. across } R_S = 0.062 \text{ volt}$$
$$\text{P.D. across } R_X = 0.076 \text{ volt}$$
$$\frac{R_X}{R_S} = \frac{V_X}{V_S}$$
$$\therefore R_X = 0.01 \times \frac{0.076}{0.062}$$
$$= 0.01226 \, \Omega$$

The value of R_X as marked on the ammeter shunt was 0.012531 ohm. This potentiometer method of measuring resistance is particularly useful for low resistance measurements.

Summary

Commercial forms of d.c. potentiometer are direct reading.

The range of a d.c. potentiometer may be extended by using a voltage ratio box.

By using a potentiometer to measure the p.d. across a standard resistor carrying current, the magnitude of the current may be determined with a high degree of accuracy.

Resistances may be compared by using a potentiometer to measure the voltage drops across the resistors when they are connected in series and carrying current.

Examples 7.2.

1. What are the disadvantages of the simple potentiometer compared with the direct reading potentiometer?

Explain why the magnitude of the voltage which may be measured by a potentiometer is limited, and how the range of a potentiometer may be extended.

2. Sketch a circuit for a potentiometer which is to be direct reading and explain carefully, with a complete diagram of connections, how this potentiometer may be applied to check the calibration of a voltmeter which has a scale range of 0–100 volts.

3. Show with the aid of circuit diagrams how a potentiometer may be used to check the calibration of an ammeter.

In a test where a standard cell with an e.m.f. of 1·0183 volt was used, the potentiometer current was adjusted until the cell e.m.f. was balanced by the voltage drop across 1·0183 m of the potentiometer wire. With an unknown current flowing through a standard resistor of 0·1 ohm the voltage drop was balanced across 1·475 m of wire. Calculate the unknown current.

4. How is a commercial d.c. potentiometer standardised? Describe, giving suggested values for all components, how the potentiometer would be arranged to calibrate (a) a 0–1 A ammeter and (b) a 0–50 V voltmeter.

5. Describe clearly how low values of resistance may be measured using a direct reading potentiometer. Compare this method with other methods of measuring resistance.

The construction and operation of moving coil instruments

Experiment 7(viii) *To examine the construction and operation of moving coil instruments*

The main features of the instrument as examined are shown in the diagram (Fig. 7.17).

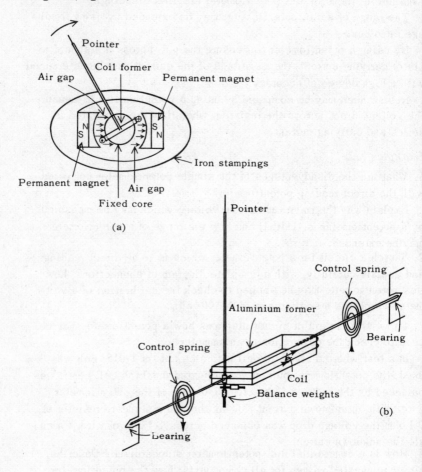

Fig. 7.17. Moving coil instrument.

A magnetic field is produced across two air gaps by rectangular Alcomax magnets (Fig. 7.17(a)). The outer iron paths of the magnetic circuit are provided by a stack of soft iron stampings. Between the air gaps there is a fixed iron core.

The moving part of the instrument (Fig. 7.17(b)) consists essentially of a rectangular coil wound on an aluminium former and mounted in jewel bearings. Two sides of the coil lie in the magnetic field produced by the magnets and current is lead in and out of the coil through fine control springs.

When current flows in the coil, forces are established on the conductors lying in the magnetic field and the torque on the moving system produces deflection of the instrument pointer. Opposing this deflecting torque is the controlling torque which the springs exert on the moving system.

The forces on the conductors of the moving coil, and hence the deflecting torque is proportional to the product of the flux density of the magnetic field and the current in the coil, Since the air gap is normally designed so that the flux density is constant over the angular range of the coil movement, the deflecting torque is proportional to the current. The controlling torque exerted by the springs is proportional to the displacement of the coil and we therefore have a uniformly divided instrument scale.

Damping

When an instrument is in use it is desirable that the pointer should move across the scale in one steady sweep and not oscillate before coming to rest in its final position. An instrument in which there is no oscillation is said to be dead beat or critically damped. In the instrument examined, damping was achieved by winding the coil on a metal former.

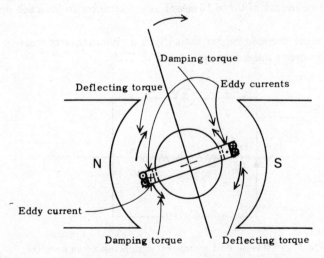

Fig. 7.18. Eddy current damping.

When the coil carries current and is moving in the magnetic field (Fig. 7.18) two sides of the coil former cut across magetic flux and e.m.f.'s are induced. A current circulates in the former and forces are produced which oppose motion. The more rapidly the coil moves, the greater is the damping effect of the eddy currents in the former.

The pointer in the instrument examined was constructed from aluminium with the tip pressed to form a knife edge, and an anti-parallax mirror under the pointer enabled readings to be taken with a high degree of accuracy.

Adjustable screw-type balance weights were provided to enable the movement to be mechanically balanced.

In a moving coil instrument temperature changes result in changes of coil resistance. This leads to instrument errors and to minimise the errors it is usual to arrange for the moving coil to be connected directly in series with a 'swamping' resistor. This is a resistor several times higher than that of the coil and with a temperature coefficient of resistance as close to zero as possible.

Extending the range of a moving coil instrument

The instrument provided was marked as giving a full-scale-deflection (*F.S.D.*) when carrying a current of 15 mA. The total resistance of the movement was 5 ohms and hence the voltage across it to give *F.S.D.* was 75 mV. The movement was therefore suitable for use as an ammeter to measure currents of up to 15 mA or as a voltmeter to indicate up to 75 mV.

To measure currents higher than 15 mA a shunt resistor must be connected across the movement as in Fig. 7.19.

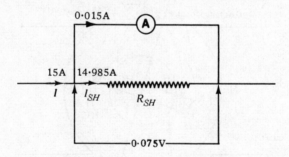

Fig. 7.19. Moving coil instrument connected as an ammeter.

If the *F.S.D.* is to be 15 A then the current through the shunt $I_{SH} = 14·985$ A and the shunt resistance,

$$R_{SH} = \frac{0·075 \text{ V}}{14·985 \text{ A}} = 0·00501 \text{ ohm}$$

This value was seen to be in agreement with that marked on a 15 ampere shunt provided for use with this instrument movement.

Figure 7.20 is a sketch of a typical ammeter shunt.

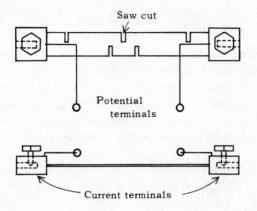

Fig. 7.20. Ammeter shunt.

Manganin strip is commonly used for ammeter shunts and the final value may be adjusted by means of saw cuts in the strip. Separate terminals are provided for connection to the main circuit and to the instrument movement. Manganin has a low temperature coefficient of resistance and also a low thermal e.m.f. with copper.

If the instrument is used as a voltmeter $F.S.D.$ will occur with a p.d. of 75 mV across the terminals. It is clear that if the movement is to be used in the measurement of higher voltages then additional resistance must be included in series with the instrument (Fig. 7.21).

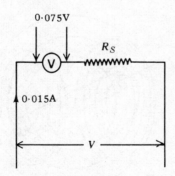

Fig. 7.21. Moving coil instrument connected as a voltmeter.

If the instrument is to give full scale deflection with 150 volts applied, then to limit the current through the movement to 15 mA the total resistance of the circuit must be given by,

$$\frac{150\,\text{V}}{0.015\,\text{A}} = 10\,000\,\text{ohms}.$$

Since the instrument movement has a resistance of 5 ohms the additional series resistor required is 9995 ohms.

When connected as a voltmeter the instrument still depends on a current through the moving coil to produce a deflecting torque. The current is proportional to the p.d. across the instrument terminals and hence the instrument when calibrated in volts has a uniformly divided scale.

The steady deflection of a moving coil instrument depends on the average value of the current through the movement and it is only suitable for use in d.c. circuits.

If it is desired to measure alternating quantities using moving coil instruments then the alternating quantity must be changed to a direct quantity. This may be achieved by using thermocouple and rectifier devices, as discussed later.

Developments in the design of moving coil instruments have been towards improving the sensitivity, increasing the length of the scale, reducing friction at the bearings and increasing the mechanical robustness.

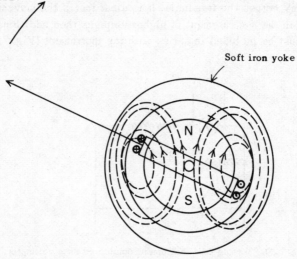

Fig. 7.22. Alternative magnetic field arrangement.

Figure 7.22 shows an arrangement where the inner core is a sintered anisotropic magnet with integral soft-iron pole shoes and an outer soft-iron yoke. The yoke serves as a screen against external magnetic fields.

Figure 7.23 shows a magnetic field arrangement which enables the movement to produce a full scale pointer deflection of 240°.

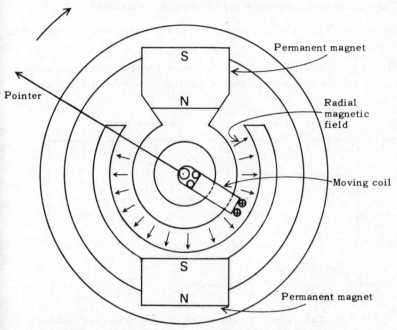

Fig. 7.23. Moving coil instrument with circular scale.

In Fig. 7.24 we show in a simplified form an alternative method of suspending the moving coil unit.

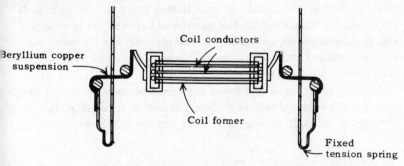

Fig. 7.24. Pivotless suspension system.

This system replaces the control springs, pivots and jewel bearings at each end of the movement, with a ribbon of beryllium copper which is held taut by fixed tension springs. In addition to supporting the movement the ribbons are used to lead current in and out of the moving coil and to provide the controlling force.

The moving coil instrument used to measure resistance

Experiment 7 (ix) *To calibrate a moving coil instrument to measure resistance*

In this experiment a moving coil instrument was connected in series with a 1·5 volt cell and a variable resistor as shown in Fig. 7.25.

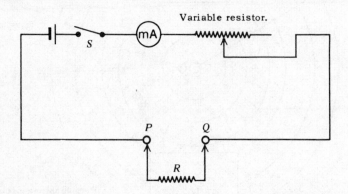

Fig. 7.25. Calibrating a moving coil instrument to measure resistance.

This circuit is similar to the circuits used in multi-range test instruments when set to measure resistance. With the terminals PQ short-circuited and the switch S closed the variable resistor is adjusted until the moving-coil milliammeter indicates full scale deflection. If additional resistance (R) is now introduced between the terminals P and Q the instrument deflection will be reduced, and by inserting various standard resistors between P and Q the instrument may be calibrated in terms of resistance.

A dial type resistance box was connected between P and Q and with the box adjusted to give different values of resistance, the readings of the milliammeter were observed and tabulated.

Resistance inserted (Ω)	0	50	120	200	300	500	700	1000	1500	2000	3000	4000	6000	10000	13000	20000
Instrument reading (mA)	1·0	0·98	0·9	0·85	0·78	0·69	0·61	0·53	0·43	0·36	0·27	0·22	0·16	0·10	0·04	0·015

Fig. 7.26 shows a replica of the instrument dial calibrated 0–1 mA together with the resistance scale.

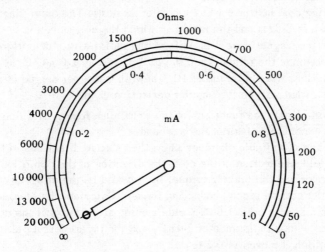

Fig. 7.26. 0–1 mA instrument calibrated to measure resistance.

It will be observed that the scale for resistance extends from zero when the current is 1 mA to ∞ when the current is reduced to zero.

Summary

Moving coil instruments may be used as voltmeters and ammeters in d.c. circuits and these instrument scales are normally evenly divided. Damping is by means of eddy currents induced in the coil former and the control torque is provided by means of hair springs.

The range of an ammeter may be extended by using instrument shunts and that of a voltmeter by connecting additional series resistors.

Moving coil instruments may be calibrated as resistance measuring instruments.

Examples 7.3.

1. Describe clearly how a deflecting torque is produced in a moving coil instrument. Why is it desirable to have a radial magnetic field in the air gap?

In a particular instrument the moving coil is square and has sides of length 20 mm. The flux density in the air gap is 0·18 T and the coil has 40 turns. Calculate the deflecting torque when the current through the coil is 1 mA.

2. What is the purpose of providing a controlling torque in a moving coil instrument? Explain what the effect would be if this control was not provided in an instrument.

A moving coil instrument has a coil of 25 turns. The axial length of the air gap is 0·02 m and the radius in which the coil moves is 0·01 m. If the magnetic field is uniform and radial with a density of 0·12 T determine the torque when the coil current is 1·0 mA. If the control springs exert a torque of 10^{-8} newton metre per degree of coil movement, what will be the angular deflection of the coil?

3. Explain clearly the functions of the aluminium former on which the coil of a moving coil instrument is wound.

By means of suitable sketches show the relative directions of the air-gap field, the motion of the coil, the direction of the e.m.f. induced in the former and the braking torque. What would be the effect of winding the coil on a non-conducting former? Calculate the braking torque on a square coil of 0·02 m side moving at 2·5 rad/s if the coil former has a loop resistance of 2·4 mΩ, and the radial density of the field in which it moves is 0·2 tesla.

4. Explain why the scale of a moving coil instrument is normally evenly divided.

A moving coil voltmeter has full scale deflection of 100 volts and its resistance is 500 ohms per volt. If the moving coil has a resistance of 5 ohms determine (a) the current for full scale deflection, (b) the additional series resistor required to limit the current, if the instrument is to be adapted to measure up to 500 volts, and (c) the voltage drop across the moving coil.

5. Describe with sketches the construction and principle of operation of a moving coil instrument if the pointer moves through an arc of about 240° at full scale deflection.

A moving coil instrument has a resistance of 5 ohms and full scale deflection is produced when the p.d. across the instrument terminals is 75 mV. Determine the values for suitable resistors which will enable the instrument to be used to measure d.c. voltage drops of up to 250 volts and currents up to 15 amperes. Give a connection diagram and devise a switching arrangement.

6. A moving coil voltmeter has a resistance of 1000 ohms. The scale is divided into 150 equal divisions. When a potential difference of 1 volt is applied to the terminals of the voltmeter a deflection of 100 divisions is obtained. Explain how the instrument could be adapted to measure up to 300 volts. Give a circuit diagram.

7. It is required to shunt a galvanometer of 420 ohms resistance so that only 10% of the main current of 5 mA shall pass through the galvanometer. Determine (a) the resistance of the shunt, (b) the combined resistance of the shunt and the galvanometer and (c) the power dissipated in the combination.

8. A moving coil milliammeter has a range of 0–10 mA and a resistance of 75 ohms independent of temperature change. It is adapted by means of switched wire wound resistors to read 0–10 volts and 0–1 ampere. The resistors are accurate at $10°C$ but the wire used in their construction has a temperature variation of 0·2% per $°C$ increase in temperature. Determine the error in each of the ranges when the room temperature is $35°C$ and the instrument pointer is indicating full scale deflection. Ignore the effect of the heat dissipated in the meter.

9. A current of 100 amperes is measured by a moving coil permanent magnet instrument fitted with a shunt. The moving coil takes 20 mA and has a resistance of 3·75 ohm at $15°C$, at which temperature the instrument reads correctly. If the coil is of copper with a temperature coefficient of 0·0042 at $0°C$ and the shunt is an alloy of negligible temperature coefficient calculate (a) the resistance of the shunt required for use at $15°C$ (b) the error of the instrument when the pointer is fully deflected and the temperature is $50°C$.

10. A multi-range meter has 2 V, 20 V and 200 V ranges and is labelled '10 000 ohms per volt'. The resistance of the movement is 80 ohms. Describe how these ranges may be achieved.

A voltmeter used to measure the p.d. of about 1·5 V between two points gives a higher numerical reading when switched to the 20 V range than when switched to the 2 V range. Can you explain this?

11. A voltmeter with a resistance of $10 k\Omega$ is used to measure the p.d. between points A and B of the circuit shown. What is the voltmeter reading? Neglect any internal resistance of the battery.

If the ammeter reading is taken while the voltmeter is connected across AB and the two instrument readings are used to calculate the value of R what is the precentage error in the value of R?

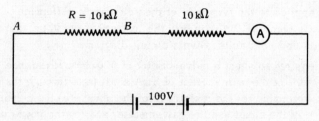

Fig. 7.27. Example 11.

12. Why is it desirable that a voltmeter should have as high a resistance as possible?

When a voltmeter was connected between points A and C in Fig. 7.28 the reading was 8 V.

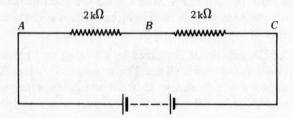

Fig. 7.28. Example 12.

Explain why the same voltmeter connected between the points A and B or B and C gave a reading of only 3 V. Calculate the resistance of the voltmeter. Ignore the battery internal resistance.

13. Two moving coil instruments give full scale deflections with currents of 10 mA and 50 μA respectively. Corresponding internal resistances are 24 Ω and 10 kΩ. One of these instruments is to be modified to read 0–500 mA and the other to read 0–1000 V. Select the more suitable instrument for each purpose and calculate the values of the additional resistors required.

14. Explain how shunts and multipliers may be used to extend the current and voltage ranges of moving coil instruments.

A 1 mA meter movement requires an 8·33 ohm shunt to extend its range to 10 mA. What is the resistance of the meter? The original meter is to be used as a multi-range meter having full scale deflections of

10 V, 100 V, and 1000 V. Draw a suitable circuit diagram including component values and a selector switch.

15. Explain carefully how a moving coil milliameter may be arranged as a resistance measuring instrument. State the additional components which are necessary and give a circuit diagram. How would you calibrate the instrument? What type of scale is associated with the ohmeter?

16. How may Ohm's law be applied to determine resistance? How could you set about determining the resistance of a resistor of about 100 ohms if provided with a milliameter of 5 ohms resistance which requires 15 mA to produce full scale deflection? Also available are resistors of 6·66 kΩ and 0·076 Ω and a variable voltage supply of 0 to 80 volts.

The thermoelectric effect

Seebeck discovered in 1821 that if two wires of different metals were joined at the ends and the junctions maintained at different temperatures then a current circulated through the wires. The two wires were said to form a thermocouple and the e.m.f. which produces the current is termed a thermoelectric e.m.f. The presence of a thermoelectric circulating current may be detected by using a galvanometer which is a sensitive moving coil instrument.

Calibrating a thermocouple

Let us carry out an experiment on a thermocouple.

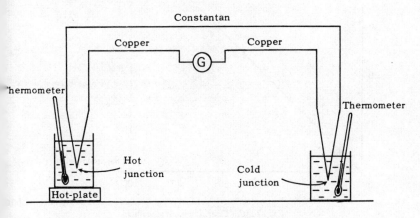

Fig. 7.29. Measuring thermoelectric e.m.f.

Experiment 7(x) *To determine the relationship between e.m.f. and temperature for a copper-constantan thermocouple*

A circuit was connected as in Fig. 7.29.

The 'hot' and 'cold' junctions were immersed in vessels containing oil. As one junction was heated the junction temperatures and galvanometer deflection were observed.

Junction hot (°C) cold (°C)	28 28	40 28	50 28	60 28	71 28	80 28	90 28	95 28
Temperature difference (°C)	0	12	22	32	43	52	62	67
Galvanometer deflection (mm)	0	4	10	15	21	24	30	32

In Fig. 7.30 the galvanometer deflection is shown plotted against the difference in temperature between the junctions.

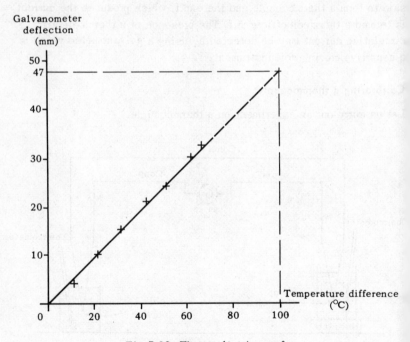

Fig. 7.30. Thermoelectric e.m.f.

From the mean slope of the graph it is seen that a galvanometer deflection of 47 mm corresponds to 100°C difference in temperature between the hot and cold junctions.

A galvanometer deflection of 1 mm thus corresponds to a temperature difference of 100°C/47 = 2·13°C.

The galvanometer constant was 0·22 μA/mm and hence 100°C temperature difference between the junctions produces a current of 47 × 0·22μA = 10·34μA.

The resistance of the galvanometer and thermocouple circuit was 420 ohms hence, the thermoelectric e.m.f. = 10·34 μA × 420 Ω
= 4350μV/100°C.

Mean thermoelectric e.m.f. = 43·5 μV/°C

Using a thermocouple to measure temperature rise

Here is an experiment to determine the temperature rise in the field windings of a d.c. machine.

Experiment 7(xi) *To measure the temperature rise at various points within the field coils of a d.c. machine*

A field coil was used in which thermocouple junctions had been embedded during the winding of the coil. Figure 7.31 shows the approximate positions of the thermocouples. The metals used were copper and constantan — similar to those used in the previous test to calibrate a thermocouple.

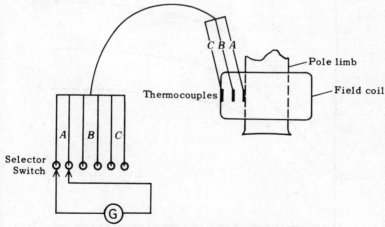

Fig. 7.31. Thermocouples embedded in a field coil.

The machine was connected as a motor and run on full load for 60 minutes. At five minute intervals during this period the deflection of a galvanometer connected across each of the thermocouples was observed. Changes in temperature were evaluated from the curve for a similar thermocouple as calibrated in the previous experiment.

Time (min)		5	10	15	20	25	30	35	40	50	60
Galvanometer deflection (mm)	A	7.5	10	12.8	14	15	16	16.5	16.5	16.9	17
	B	10	13.4	16	18	19.4	20	20.8	21	21.6	22
	C	6	8.2	10	11	12	12.5	13	13	13.2	13.5
Temperature rise (°C)	A	16	21.3	27.2	29.8	31.9	34	35	35	36	36.2
	B	21.3	28.5	34	38.2	41.2	42.5	44.2	44.6	46	46.8
	C	12.8	17.5	21.3	23.4	25.5	26.6	27.6	27.6	28	28.7

The curves (Fig. 7.32) show the temperature rise plotted against time. It will be observed that the highest temperatures are reached inside the winding.

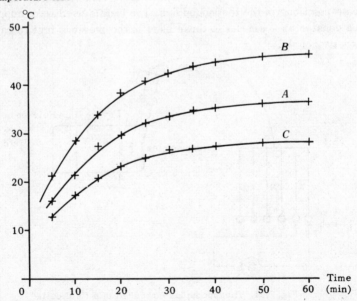

Fig. 7.32. Temperature rise of a field coil as determined by using thermocouples.

Thermocouple ammeter

A thermocouple is a very convenient device for measuring temperatures in positions where it is not possible to attach thermometers — such as within coils. The principle of the generation of a thermoelectric e.m.f. may also however be used to measure current.

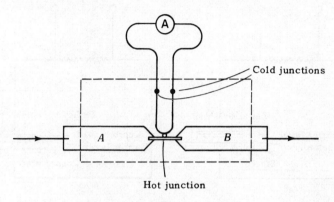

Fig. 7.33. Thermocouple ammeter.

In one form of thermocouple ammeter the current to be measured flows through a platinum alloy resistor which heats one junction of a copper-constantan thermocouple (Fig. 7.33). The cold ends of the thermocouple are soldered to copper strips which are in thermal but not electrical contact with the heavy terminal strips A and B of the instrument.

This instrument may be calibrated with a direct current and will then measure the r.m.s. values of alternating currents having any waveform or frequency.

Instrument rectifiers

It was stated earlier that moving coil instruments could be used in conjunction with rectifiers to measure alternating quantities.

Rectifiers most commonly used are of the selenium or copper oxide types as shown diagrammatically in Fig. 7.34. Rectifiers have the common property of permitting a current to flow with comparatively low resistance in one direction but they offer a high resistance to current flow in the reverse direction.

In the selenium rectifier (Fig. 7.34(a)) a barrier layer is formed by suitable heat treatment which allows electrons to pass readily from the tin alloy counter electrode to the selenium but the resistance to

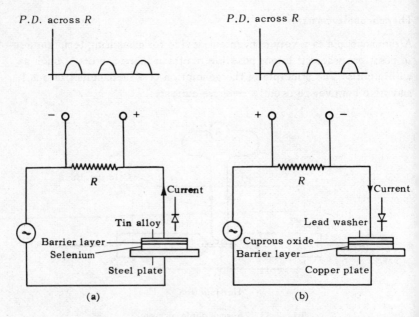

Fig. 7.34. Selenium and copper oxide rectifiers.

the movement of electrons in the reverse direction is very high. Thus if an alternating voltage is connected in series with such a rectifier then the voltage developed across a resistor (R) in the circuit will be unidirectional.

In the copper oxide rectifier (Fig. 7.34(b)) electrons pass freely from a copper plate into a layer of cuprous oxide on one side of the plate but the resistance to movement in the reverse direction is very high.

Symbols shown alongside the rectifiers in Fig. 7.34 are those commonly used to represent rectifiers. The current flows with very little resistance in the direction of the apex of the triangle.

When a rectifier is used in an a.c. circuit it must be able to withstand the maximum inverse voltage of the supply during the negative half cycles. When dealing with other than very low voltages it is usual to connect a number of units in series to ensure that breakdown does not occur.

Germanium and silicon rectifiers which have been developed are capable of dealing with very heavy current outputs and high reverse voltages. Solid state rectifiers are used in power circuits where single or polyphase a.c. supplies are to be converted to supply d.c. requirements. At this stage we will concern ourselves only with a test carried out on a single rectifier unit intended for use with a moving coil instrument.

The characteristic curve for a rectifier unit

Here is an experiment carried out to determine the operating characteristic for an instrument rectifier.

Experiment 7(xii) *To plot the characteristic curve for an instrument rectifier*

The circuit used was connected as in Fig. 7.35.

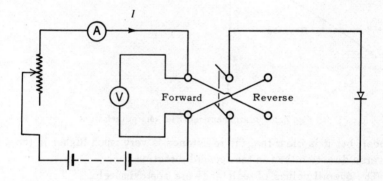

Fig. 7.35. Circuit used to determine rectifier characteristics.

The voltage applied to the rectifier unit was increased from zero to a maximum of 0·36 volts in the forward direction and the resulting currents observed. An increasing voltage was then applied in the reverse direction. Observations were as tabulated.

Forward direction						
V	(V)	0	0·1	0·2	0·3	0·36
I	(mA)	0	0·02	0·14	0·6	1·0
Reverse direction.						
V	(V)	0	1	2	3	
I	(μA)	0	1	2·3	3·8	

It will be observed that different scales have been used for the forward and reverse currents in plotting the rectifier characteristic (Fig. 7.36). The resistances in the forward and reverse directions are not

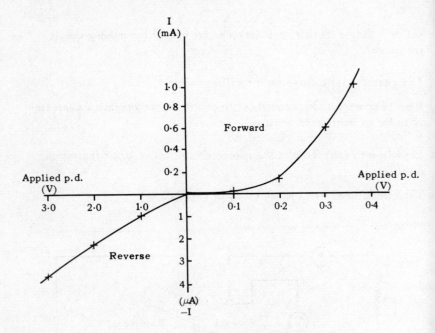

Fig. 7.36. Instrument-rectifier characteristics.

linear, but it is clear that the resistance is very much higher in the reverse direction than in the forward direction.

The overall values of resistance are approximately,

$$\frac{0.36 \text{ V}}{0.001 \text{ A}} = 360 \text{ ohms in the forward direction}$$

and

$$\frac{3 \text{ V}}{3.8 \,\mu\text{A}} = 0.79 \times 10^6 \text{ ohms in the reverse direction}$$

Let us now consider how four rectifier units may be connected to form a bridge network for use with a moving coil instrument.

Full wave rectifier for use with a moving coil instrument

Experiment 7 (xiii) *To examine a rectifier for use with a moving coil instrument*

The unit examined was a full wave rectifier unit. It was intended for use with an indicator movement of 100 ohms resistance giving a full scale deflection of 100 divisions when carrying a current of 1 mA. The rectifier unit enabled the instrument to be used as an a.c. voltmeter with ranges of 0–10 V, 0–100 V, and 0–500 V.

Figure 7.37 shows the connections.

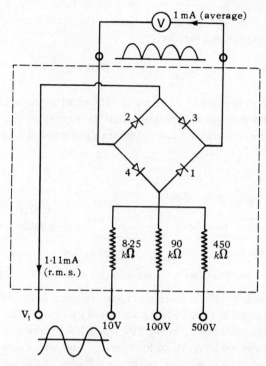

Fig. 7.37. Moving coil instrument and rectifier unit.

Let us assume that an alternating voltage is connected across V_1 and the 10V terminal. It will be seen that when the 10V terminal is positive with respect to V_1, current flows through rectifier 1 and then through the instrument movement before returning through rectifier 2 to the supply. During the intervening half cycles when the 10V terminal is negative with respect to V_1, the current flow is through rectifier 3 to the instrument and then back through rectifier 4 to the 10V terminal. The current through the instrument is thus a pulsating direct current and we refer to this rectification, where both halves of the a.c. wave pass through the instrument, as full wave rectification.

The deflection of the instrument moving coil movement depends on the average current through it. It is usual to calibrate an instrument for use with a rectifier in terms of r.m.s. values, assuming a sinusoidal wave with a form factor of 1·11.

For full scale deflection to correspond to a p.d. of 10 V (r.m.s.) between V_1 and the 10 V terminals the average value of the pulsating current through the instrument must be 1 mA and the r.m.s. value of the supply current will be $1 \times 1 \cdot 11 = 1 \cdot 11$ mA. The total resistance presented to the supply is given by,

$$\frac{V}{I} = \frac{10}{1 \cdot 11} \times 10^3 = 9000 \text{ ohms}$$

On the 10 V range this is made up of 8250 ohms in the series resistor, 100 ohms in the instrument movement, and the remaining 650 ohms in the two rectifier units which are conducting in series at any instance (Fig. 7.38).

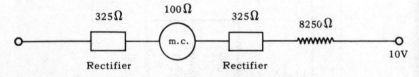

Fig. 7.38. Rectifier instrument circuit on the 10 V range.

The non-linearity of the rectifier characteristics in the forward direction has negligible effect when considered along with the series resistor of over 8000 ohms. On the 100 V and 150 V ranges it will be observed that the resistances of the rectifier and instrument movement have been neglected in relation to the magnitudes of the series resistors of 90 kΩ and 450 kΩ respectively.

If the instrument movement is disconnected from the rectifier unit, then when a p.d. is applied across the input to the rectifier there will be no current through the series multiplier resistors. The full voltage will appear across the rectifier and this may lead to breakdown of the unit. It will be observed that the resistance of the instrument is about 900 ohms per volt and that the average current for full scale deflection is 1 mA.

Rectifier voltmeters may be designed with very high resistance values. They have low power consumption and are not affected by frequency changes within wide limits.

Summary

Temperature measurements may be made using a calibrated thermocouple with a sensitive moving coil instrument.

A thermocouple ammeter indicates r.m.s. values whatever the frequency and waveform of the current being measured.

Instrument rectifiers when used in conjunction with moving coil instruments enable these instruments to be used to measure alternating voltages.

Examples 7.4.

1. What is meant by a thermoelectric e.m.f.?

Describe an experiment to determine the e.m.f./temperature relationship for a thermocouple. When is it preferable to measure temperature using a thermocouple rather than a thermometer?

2. Sketch two types of rectifier and show a typical characteristic for a solid contact rectifier.

3. Describe the principle of the thermocouple milliameter. What are its advantages and limitations in the measurement of high frequency currents?

A thermocouple ammeter is connected in circuits carrying (a) 5 A d.c., (b) a sinusoidal current and (c) a current of square waveform. If the instrument indicates the same in each circuit calculate the maximum value of the alternating c urrents.

4. Compare the characteristics of a.c. milliameters working in conjunction with (a) full-wave bridge-type metal rectifiers and (b) thermocouples. If both instruments are calibrated on a 100 Hz supply which has a sinusoidal wave form how will their readings compare when they are connected to a 100 Hz source having a rectangular waveform?

5. Explain the principle of a moving coil ammeter for a.c. or d.c. use giving details of (a) the arrangement of the pole pieces, (b) a rectification circuit for a.c. voltmeter use, and (c) the extension of the instrument ranges.

A d.c. meter of resistance 5 ohms gives a full scale deflection when he current is 15 mA. Design a shunt to give ranges of 0–0.15 A, 0–1.5 A and 0–15 A.

Moving iron instruments

Experiment 7(xiv) *To examine the construction and operation of a moving iron instrument*

The moving iron instrument examined consisted of a coil in which was mounted a fixed iron. A second iron was attached by an arm to a spindle passing through the centre of the coil (Fig. 7.39).

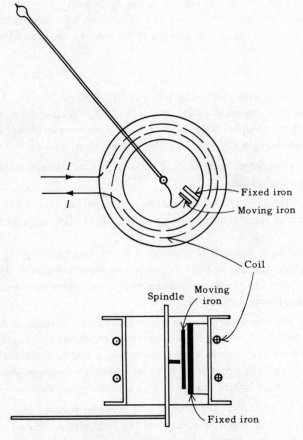

Fig. 7.39. Moving iron instrument.

When current flows through the coil the fixed and moving irons become similarly magnetised and they repel one another. A pointer which forms part of the moving system deflects to the right.

The instantaneous force of repulsion acting between the fixed and moving irons depends on the square of the instantaneous current. Thus the mean deflecting torque is proportional to the average value of the (current)^2. The instrument gives an indication of the r.m.s. value of the current through the coil. It may be used in a.c. and d.c. circuits.

The deflecting torque is opposed by a control torque produced by hair springs as in a moving coil instrument. The instrument has an unevenly divided scale because the deflecting torque follows a square law. By suitably shaping the fixed and moving irons modifications to the scale divisions may be produced. Damping is by mechanical means. In one arrangement vane(s) move in an air-tight box. Part of a typical damping sustem is shown in Fig. 7.40 where the cover of the box has been removed.

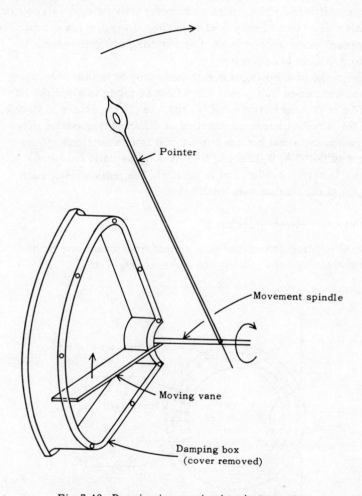

Fig. 7.40. Damping in a moving iron instrument.

When connected in d.c. circuits the instrument may show higher readings with decreasing currents than with increasing currents. This is caused by the magnetic hysteresis effect in the fixed and moving irons.

In a.c. circuits errors may be introduced with changes in frequency which vary the reactance of the coil. Precautions must be taken to avoid eddy currents being established in adjacent metal.

When a moving iron instrument is intended for use as a voltmeter the coil consists of many turns of wire with a comparatively high resistance. If it is to be used as an ammeter then the coil will consist of only a few turns of wire or strip having a very low resistance. The instrument examined was marked as requiring 210 ampere turns to produce full scale deflection.

The range of a moving iron instrument may be extended by using shunts and series multipliers. Care must be taken to avoid the introduction of thermoelectric e.m.f.'s, and when using shunts to ensure that the impedance remains constant at different frequencies. The instrument examined had the current coil in four sections, giving ranges of 0–7·5 A, 0–15 A and 0–30 A with the coils connected in series, in series-parallel and in parallel. The resistance of each section of the winding was 0·0068 ohm.

The use of a current transformer

In an alternating current circuit a current transformer may be used to extend the range of a moving iron ammeter (Fig. 7.41).

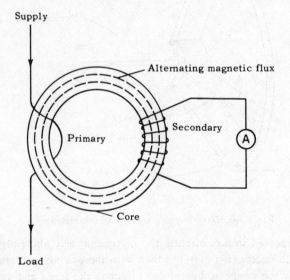

Fig. 7.41. Connection of a current transformer.

The primary winding of the current transformer consists of one or more turns. It carries the current to be measured and there is a negligible voltage drop across the primary winding.

An ammeter of suitable range is connected across the terminals of the transformer secondary winding. The alternating core flux depends upon the primary current and as in any transformer the primary and secondary ampere turns must balance. The primary current in a current transformer is the load current and is not influenced by the transformer secondary current.

If the secondary circuit of a current transformer is open circuited while the primary circuit is carrying load current, then with no secondary ampere turns the core flux will reach a very high value. This results in very high voltages being induced in the transformer secondary winding and probable breakdown of the transformer insulation.

The electrodynamic wattmeter

Experiment 7(xv) *To examine the construction and principle on which the electrodynamic or dynamometer type wattmeter operates*

The circuits of the instrument examined were basically as shown in Fig. 7.42.

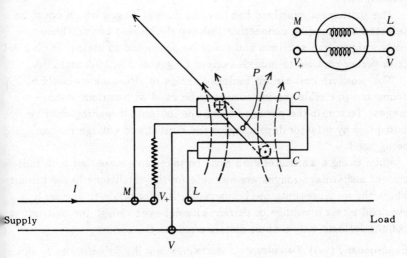

Fig. 7.42. Dynamometer type wattmeter.

The two coils C are fixed and establish a magnetic field which depends on the current in the circuit in which the wattmeter is connected.

The moving coil P is mounted in the magnetic field produced by the coils C. It is connected in series with a high resistor and carries a current proportional to the voltage applied across the load.

The deflection of the instrument pointer depends on the torque which is produced on the moving coil. This torque is proportional to both the current in the moving coil and the strength of the magnetic field in which it lies. Hence in a d.c. circuit the deflecting torque is proportional to the product of the volts and the amperes — or the power in the circuit. Fine springs provide the controlling torque and are used to lead the current in and out of the moving coil. An almost linear scale is usual, although by shaping the fixed coils to alter their field distribution the scale law may be modified.

The instantaneous deflecting torque in an a.c. circuit is proportional to the product of the instantaneous values of current and voltage. The deflection produced is proportional to the mean deflecting torque and hence in an a.c. circuit the instrument indicates the mean power, $V \times I \times \cos \phi$.

Damping is achieved mechanically by vanes moving in a box, as in the moving iron instrument. The instrument is vulnerable to stray magnetic fields and the movement is screened with a metal of high permeability.

The instrument examined had several current ranges which could be selected by changing connecting links to the current coils. These coils were in four sections and could be connected in series, in parallel or in series-parallel to provide current ranges of 5 A, 10 A and 20 A.

The pressure coil circuit included series reisitors which could be connected to enable the instrument to be used with various voltage ranges. To obtain the power in watts the instrument reading must be multiplied by a factor depending on the current and voltage ranges being used.

When using a wattmeter care must be taken to ensure that both the current and voltage ranges are suitable for the conditions in the circuit where the measurements are being made. It is possible to seriously overload either a voltage or current element even though the instrument pointer deflects only a short distance across the scale.

Experiment 7(xvi) *To connect a wattmeter in a d.c. circuit and to make corrections for power losses in the wattmeter elements*

In Fig. 7.43. the resistor to the right of the line $X-X$ constitutes a load and the voltmeter across the resistor will be considered to form part of this load.

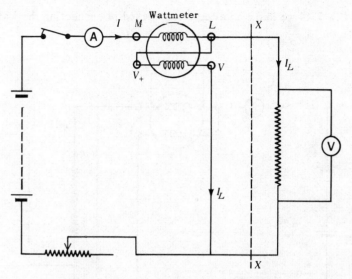

Fig. 7.43. Connections to a wattmeter.

The current through the wattmeter current coil comprises the load current I_L and the current I_V through the wattmeter voltage element,

$$I = I_L + I_V$$

Power indicated by the wattmeter $= V(I_L + I_V)$
$\qquad\qquad\qquad\qquad\qquad\;\; =$ load power + power in the wattmeter voltage element.

In the test carried out the variable resistor in the circuit was adjusted until the voltage drop across the load was 80 volts. The instrument readings were as tabulated.

Load	Supply current	Wattmeter reading	Wattmeter constant	Power
(V)	(A)	(W)		(W)
80	1·41	7·1	16	113·6

Wattmeter voltage element resistance (R_V) = 1200 ohms

Voltage across voltage element = 80 volts

Power in wattmeter voltage element $= \dfrac{V^2}{R_V}$

$$= \frac{80^2}{1200} = 5.33 \text{ watts.}$$

Power in load = 113·6 − 5·33 = 108·27 watts.

In Fig. 7.44 we show an alternative method of connecting the wattmeter.

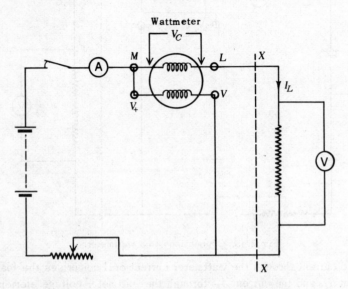

Fig. 7.44. Alternative wattmeter connections.

The current through the wattmeter current element is now the true load current I_L but the voltage drop across the pressure circuit element is now the voltage drop V across both the load and the wattmeter current element.

$$V = V_L + V_C$$

Power indicated by the wattmeter $= I(V_L + V_C)$

= Load power + power in wattmeter current element.

In a circuit with a wattmeter connected as in Fig. 7.44 the variable resistor was adjusted until the voltage across the load was 80 volts. The instrument readings were tabulated.

Load (V)	Supply current (A)	Wattmeter reading	Wattmeter constant	Power (W)
80	1·41	6·75	16	108

Wattmeter current element resistance (R_C) = 0·365 ohm

Wattmeter voltage element resistance (R_V) = 1200 ohms

$$\text{Current through wattmeter current coil} = 1\cdot41 - \frac{80}{1200}\text{A}$$
$$= 1\cdot41 - 0\cdot067 \text{ A}$$
$$= 1\cdot343 \text{ A}$$

Power in the wattmeter current element = $I^2 R_C$

$$= 1\cdot343^2 \times 0\cdot365 = 0\cdot655 \text{ W}$$

Power in the load = 108 − 0·655 W
$$= 107\cdot35 \text{ W}$$

To confirm the value for the power as determined using the wattmeter connections a circuit incorporating a voltmeter and ammeter was connected as in Fig. 7.45.

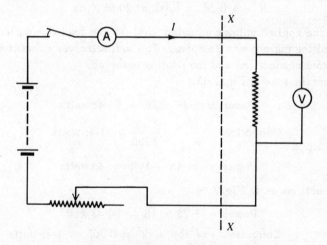

Fig. 7.45. Using a voltmeter and ammeter to measure power.

The voltmeter is considered as being part of the load and the variable resistor was adjusted until the voltage drop across the load was 80 volts. It was observed that the ammeter indicated 1·34 A.

$$\text{Power} = V \times I$$
$$= 80 \times 1\cdot34 = 107\cdot2 \text{ W}$$

It will be noted that providing the appropriate corrections are made to the wattmeter readings, both methods of connection give a result for the power in the load very close to the value obtained from the ammeter and voltmeter indications. It will also be observed that in this particular experiment the power loss in the wattmeter voltage element is higher than that in the current element, and that the second method of connection requires a smaller correction.

Can we determine the load for which the two losses in the elements will be equal?

For equal losses,

$$\frac{V^2}{R_V} = I^2 R_C$$

$$\therefore \text{load resistance, } R = \frac{V}{I} = \sqrt{R_C R_V}$$

and for the wattmeter used this corresponds to a load of

$$R = \sqrt{0.365 \times 1200} = 20.95 \text{ ohms.}$$

With the applied voltage adjusted to 41·9 V the load was varied until the resulting current was 2 amperes. The wattmeter was connected with alternative connections and the reading observed.

Connections as in Fig. 7.43.

$$\text{Power} = 5\cdot 28 \times 16 = 84\cdot 48 \text{ watts}$$

$$\text{Correction} = \frac{V^2}{R_V} = \frac{41\cdot 9^2}{1200} = 1\cdot 46 \text{ watts}$$

$$\text{Power} = 84\cdot 48 - 1\cdot 46 = 83 \text{ watts}$$

Connections as in Fig. 7.44.

$$\text{Power} = 5\cdot 28 \times 16 = 84\cdot 48 \text{ watts}$$

$$\text{Correction} = I^2 R_C = 2^2 \times 0\cdot 365 = 1\cdot 46 \text{ watts}$$

$$\text{Power} = 84\cdot 48 - 1\cdot 46 = 83 \text{ watts.}$$

It is seen that as expected the wattmeter indication is the same for both methods of connection and equal corrections are required.

Electrodynamic voltmeters and ammeters

The electrodynamic instrument is generally used as a wattmeter, but if the coils are wound with fine wire and connected in series (Fig. 7.46) the instrument may be used as a voltmeter to measure r.m.s. voltages.

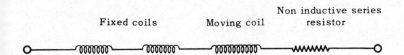

Fig. 7.46. Internal connections to dynamometer voltmeter.

The instrument may also be used as an ammeter. In this case the moving coil is connected in series with a swamping resistor across the fixed coils and a shunt resistor as in Fig. 7.47.

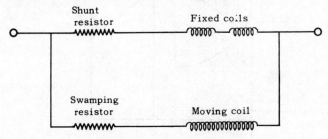

Fig. 7.47. Internal connections of a dynamometer ammeter.

Since the instruments are air-cored they are not affected by variations in frequency and wave form. They may be used in alternating circuits for measurements work where a high degree of accuracy is essential. Electrodynamic instruments tend to be more expensive and introduce higher power losses into circuits than most other types of instrument.

The ranges of dynamometer instruments may be extended by using current and/or potential transformers.

Electrostatic voltmeters

The electrical indicating instruments we have already considered depend for their action, whether voltmeter or ammeter, on the flow of a current through them. It is desirable that a voltmeter should have as high a resistance as possible — otherwise when connected across a component the shunting effect may be considerable and disturb the conditions in the circuit.

It has been seen in Volume 1 Chapter 7 that a force of attraction exists between two oppositely charged parallel plates, and the principle of the electrostatic voltmeter is based upon this force.

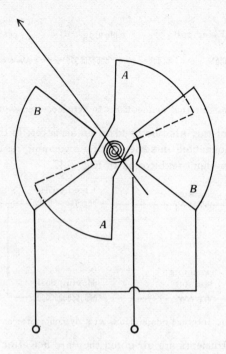

Fig. 7.48. Electrostatic voltmeter.

In the quadrant form of instrument shown in Fig. 7.48 when a voltage is applied across the two sets of plates the moving vane $A-A$ is attracted towards the fixed vane $B-B$. The control torque is provided by springs and damping by means of a damping vane.

The scale follows a fundamental square law but may be modified by shaping the vanes. The instrument can be used to measure alternating or direct voltages. In high frequency circuits errors may be introduced by capacitance currents.

The range of the instrument is extended by using a series capacitor or by connecting the instrument across one unit of a capacitor voltage multiplier.

The fluxmeter

Although the theory of the fluxmeter used in our earlier experiments is beyond the scope of this present volume the instrument is essentially a moving coil instrument with heavy electromagnetic eddy-current damping. The moving coil is suspended by a fibre in a magnetic field produced by permanent magnets. Current is lead in and out of the coil through annealed silver strips. The controlling torque is reduced to the minimum possible value.

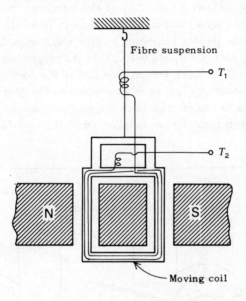

Fig. 7.49. Fluxmeter coil.

It may be shown that the angular deflection of the coil is proportional to the change in flux through a search coil connected across the terminals T_1 and T_2. In use the coil is returned to the zero position by a mechanical device before a reading is taken. Providing the radial magnetic field is uniform the scale is evenly divided and calibrated in terms of weber turns.

Insulation resistance measuring instruments

Experiment 7(xvii) *To examine an insulation resistance measuring instrument*

The simplified circuit diagram of this instrument as commonly used to measure high resistance is shown in Fig. 7.50.

The instrument examined incorporated a small permanent magnet hand-driven d.c. generator with a clutch which slipped at a predetermined speed so that a steady e.m.f. was generated. The two coils of the instrument movement are rigidly fixed at an angle to one another, and currents are lead in and out of the coils through ligaments (l).

The unknown resistor is connected between the line and earth terminals. If the resistance under test is infinitely high no current will flow through the deflecting coil. Current flows through the control coil and this coil takes up a position with its axis in line with the main magnetic field, the pointer indicating infinite resistance. If the resistance undergoing test has a zero value, then the torque produced by the deflecting coil will be much greater than that of the control coil and the instrument pointer swings across the scale to the zero position. For intermediate values of resistance the pointer takes up a position where the deflecting coil torque is balanced by the control coil torque.

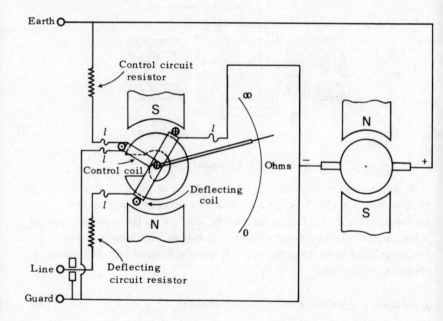

Fig. 7.50. Insulation resistance testing instrument.

The output from the internal generator was 500 volts but instruments of this type are manufactured which test up to 2500 volts.

Summary

Moving iron instruments may be used in both a.c. and d.c. circuits. Instrument scales are usually unevenly divided. Damping is by mechanical means and hair springs provide the control torque.

The range of a moving iron voltmeter may be extended by connecting additional series resistance. The range of an ammeter may be extended by connecting various sections of the coil in series or parallel arrangements. Alternatively, shunts may be used, and in a.c. circuits current transformers connected.

The secondary winding of a current transformer must never be open circuited when the primary winding carries current.

Electrodynamic instruments may be used as wattmeters in a.c. and d.c. circuits. Corrections should be made for internal power losses in the instrument elements.

Electrostatic voltmeters do not take any currents when connected in d.c. circuits and only a negligible current in a.c. circuits.

Insulation-resistance measuring instruments enable us to measure insulation resistance whilst a test voltage is being applied to the circuit.

Examples 7.5.

1. Describe the construction and principle of operation of a moving iron voltmeter. How is damping achieved and what type of scale is usually associated with this instrument?

2. How may the range of a moving iron ammeter be extended using a current transformer? What precautions must be observed when using current transformers? What is the effect of short circuiting the secondary winding of a current transformer?

3. Describe the construction and operation of an electrodynamic type wattmeter. How may different ranges be obtained using a single instrument? What is the purpose of providing damping in this instrument and how is it applied?

4. Explain clearly the meaning of the statement 'an error is introduced into power measurement when using an electrodynamic wattmeter depending on how the pressure coil circuit is connected.' Illustrate your answer with circuit diagrams and show how allowances may be made to correct these errors.

5. What advantages are to be gained by using air-cored electrodynamic voltmeters and ammeters in preference to other types of instruments? Explain the principle of action of the electrodynamic instruments and include circuit diagrams.

6. Describe the principle of the electrostatic voltmeter. What advantages has it compared with other types of voltmeter? How may the range be extended?

7. How does an insulation resistance measuring instrument operate? Sketch a circuit diagram.

A 200 metre length of cable has an insulation resistance between the core and sheath of 0·9 MΩ. Estimate the insulation resistance of a 100 metre length of similar cable.

8. Electrons in Thermionic Valves

Introduction

Since a current of electricity flowing in a circuit may be looked upon as a controlled drift of electrons every electrical circuit is an 'electronic' circuit. However the word 'electronic' has come to be associated to a very large extent with communications and control circuits.

In this chapter you will be introduced to the fundamentals of 'electronics' and with an understanding of the basic concepts you should have no difficulty in proceeding with further studies.

The diode valve

Thomas Alvar Edison carried out many experiments while developing the incandescent electric lamp and in 1878 he observed that when a carbon filament lamp had been in use for some time the inner surface of the glass bulb become coated with a dark deposit. He deduced that this was caused by particles emmitted from the heated filament settling on the glass. In attempting to prevent this darkening he investigated the effect of inserting a metal plate between the filament and glass bulb (Fig.8.1).

To his amazement Edison discovered that when a positive potential was applied to this plate an electric current flowed between the plate and the lamp filament. This is known as the Edison effect. Sir Ambrose Fleming followed up Edison's observations and in 1904 produced the first thermionic diode valve. This consisted of a carbon filament lamp with a nickel plate mounted within the loop of the filament and connected to a wire sealed through the side of the bulb.

Let us carry out an experiment on a diode valve.

Experiment 8(i) *To determine the operating characteristics for a diode valve*

A circuit was connected as in Fig.8.2.

295

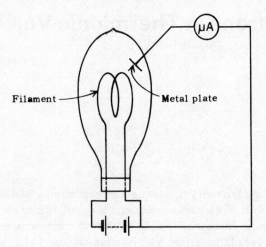

Fig. 8.1. Illustrating the Edison effect.

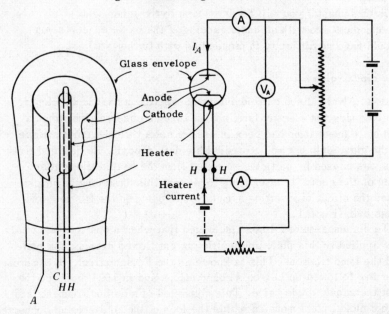

Fig. 8.2. Diode valve and circuit used in determining the valve characteristics.

The thermionic valve consists of an evacuated glass envelope and includes a filament heated by passing a current through it. Surrounding the filament is a nickel tube coated with oxides such as barium or strontium which readily emit electrons when heated. This electrode is termed the cathode. The plate or anode takes the form of a hollow metal box surrounding the cathode and spaced a small distance from it. Both electrodes are rigidly supported.

The filament was connected to a 6 volt supply and the current through it adjusted to the normal operating value of 1·3 amperes. The heating effect of this current produced a rise in the temperatures of the cathode. A variable direct voltage (V_A) was applied between the anode and cathode terminals of the valve — the anode being maintained at a positive potential with respect to the cathode. The anode voltage was increased in steps and the resulting anode current (I_A) observed.

The experiment was repeated with the current through the filament reduced to 0·9 ampere and then to 0·8 ampere. Operating a coated cathode at a temperature lower than normal can be detrimental to a cathode but is justified in this particular experiment where we wish to demonstrate the resulting effect on the valve characteristics.

Observations of the instrument indications were as tabulated.

Filament (heater) current; 1·3 A

V_A (V)	20	40	60	80	100	115
I_A (mA)	8	22	39	57	79	95

Filament current: 0·9 A

V_A (V)	20	40	60	80	100	120
I_A (mA)	3·5	13	26	42	59	76

Filament current: 0·8 A

V_A (V)	20	40	60	80	100	120
I_A (mA)	1·5	8	16	21·5	22·5	24

These observations are shown plotted in the curves of Fig.8.3.

The effect of reversing the connections in the anode circuit, so that the anode was negative with respect to the cathode was also investigated. It was observed that no anode current flowed with increasing values of negative anode voltage. It is because of this ability to conduct in only one direction that the device is termed a valve. You may care to compare this diode thermionic valve which only allows electrons to pass in one direction with the valve in the inner tube of a bicycle tyre which allows air to enter the tube from a pump during inflation but does not allow it to return when the pump is removed.

Now, how can we account for the flow of current in the anode circuit, and the particular form that the characteristics take?

Let us consider the diode valve with no voltage applied to the anode.

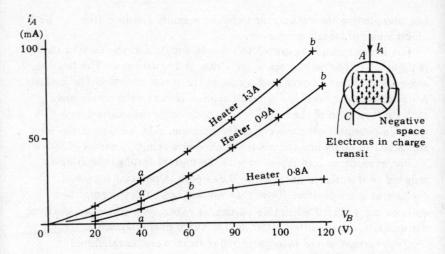

Fig.8.3. Diode valve characteristics and 'space charge'.

When a metal is heated it becomes capable of emitting electrons. Since the cathode in a thermionic valve is heated when current flows through the valve heater the cathode can emit electrons. When any electrons are emitted the cathode has an excess of positive charges and it therefore exerts a force on the escaping negative electrons attracting them back towards it.

When a positive voltage is applied to the anode, the anode also exerts a force of attraction on electrons liberated from the hot cathode. If the anode force on the liberated electrons is greater than the force exerted by the cathode, then there is a drift of electrons within the valve from the cathode to the anode. From the anode the electrons pass through the miliammeter of the external circuit and return to the cathode. We are thus able to explain the presence of the anode current. The conventional direction of current flow through the circuit is opposite to that of the electron movement. Current flows in the anode circuit as represented by I_A in Fig.8.2.

Let us look at the characteristics in Fig.8.3. The anode current increases very slightly with increasing small anode voltages. To produce a current, the force resulting from the potential applied to the anode, must overcome the force of attraction which the cathode exerts on the electrons in transit form the anode to the cathode. Electrons in transit form a negative space charge round the cathode which tends to repel any further electrons leaving the cathode. As the anode voltage is increased the anode force of attraction overcomes the effect of this space charge and the electron current through the valve increases.

The part of the characteristic a–b is comparatively straight.

If the anode voltage is increased beyond the point 'b' the curves begin to flatten and we say that the valve is becoming saturated. This saturation effect is only evident in the characteristic plotted for a heater current of 0·8 ampere. The rate at which the cathode is capable of emitting electrons depends on its temperature and when the anode voltage is made sufficiently high to collect all the electrons which the cathode is capable of emitting then any further increase in the anode voltage does not produce a higher anode current. The point at which saturation begins to occur thus depends on the heater current and in the tests carried out with heater currents of 1·3 A and 0·9 A the onset of saturation is not evident. To obtain saturation with these higher heater currents much higher anode voltages would have to be applied.

The use of the diode valve as a rectifier

You have seen that current will only flow through a diode valve when the anode is positive with respect to the cathode. Here is an experiment in which an alternating voltage is applied between the anode and cathode of a diode valve.

Experiment 8 (ii) *To show how a diode valve may be used to produce rectification of an alternating voltage supply*

In the first part of this experiment a circuit was connected as in Fig.8.4. The cathode of the valve was heated by passing an alternating current through it. This current was supplied from a low-voltage heater winding on a transformer.

The alternating voltage applied to the transformer was examined with a cathode-ray oscilloscope (C.R.O.) and observed to be sinusoidal. The output voltage across a load resistor, connected in series with the diode valve across the secondary winding of the transformer, was seen to be uni-directional but of a pulsating nature. The diode valve only allows current to pass through it when end A of the transformer secondary winding is positive with respect to end B. Hence current only flows in the secondary circuit during alternate half cycles of the applied voltage. This is known as half-wave rectification. The diode valve performs a function similar to that of a single solid state copper oxide or selenium rectifier which we considered when dealing with moving coil instruments applied to a.c. circuits (Chapter 7).

By connecting a capacitor across the load terminals (Fig.8.5) it was observed that a much steadier output voltage was obtained.

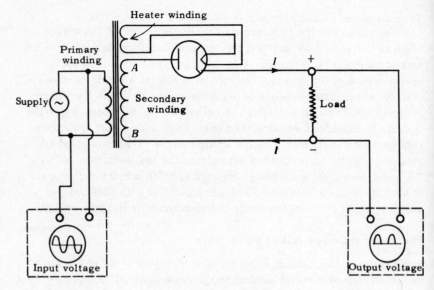

Fig.8.4. Half-wave rectification.

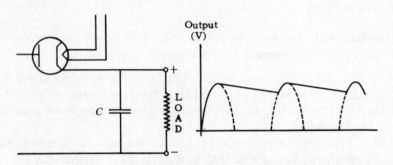

Fig.8.5. Effect of including a reservoir capacitor.

The capacitor acquires a charge when the valve is conducting and releases this charge during the half-cycles when the valve is not conducting.

In a further experiment a double-diode valve was used. This valve included two anodes and one cathode assembled in a single evacuated glass envelope.

Experiment 8 (iii) *To connect a double diode valve as a full wave rectifier*

The valve was connected to a centre tapped transformer as shown in

Fig.8.6. The output voltage which appeared across the load resistor was investigated on an oscilloscope.

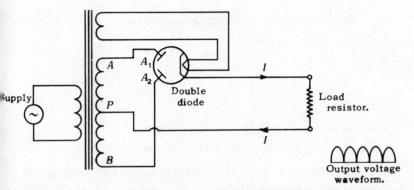

Fig.8.6. Full wave rectification.

With the connections shown one of the points A or B is always positive with respect to the centre tapping P of the transformer secondary winding. Hence the valve will always conduct through one of the anodes A_1 or A_2, and the output voltage is a fully rectified wave. A diode or double diode is commonly used to provide a d.c. supply in electronics equipment which is to be operated from a.c. mains.

By including a reservoir capacitor C across the load considerable smoothing of the output voltage was obtained. The improved voltage wave is shown in Fig.8.7(a).

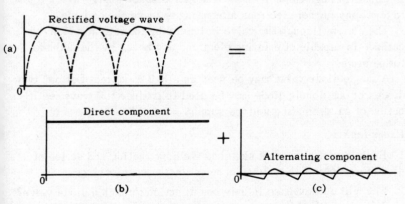

Fig.8.7. Components of a rectified output wave.

The voltage in Fig.8.7(a) is composed of a steady direct component (b) and an alternating component (c). A smoothing circuit may be used

301

as in Fig.8.8. to separate the components of the output wave.

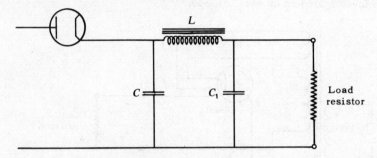

Fig.8.8. Additional $L-C_1$ smoothing circuit.

By a suitable choice of L and C_1 the alternating component of the load current may be diverted through C_1 and only the direct component of the output voltage appears across the load.

Values of C_1 and L must be chosen to make X_C small with respect to X_L in so far as the alternating component of the load current is concerned. Typical values might be 10–30 H for L and 16–32 μF for C. C and C_1 are usually electrolytic capacitors.

Summary

Electrons emitted by a hot cathode in a diode valve are attracted towards the anode when a positive potential is applied. The electrons in transit through the valve form a negative space charge which tends to repel any further electrons attempting to leave the cathode.

The current through the valve is limited by the rate at which the cathode is capable of emitting electrons and hence by the cathode temperature.

A single diode valve may be used as a half wave rectifier and two diodes or one double diode may be used to produce full wave rectification of an alternating voltage supply.

Examples 8.1.

1. Explain with the aid of sketches the construction and action of a diode thermionic valve. What is meant by the space charge?

2. Why will current pass in only one direction through a diode valve? Describe with sketches an experiment to determine the static characteristics of a diode valve with normal current through the heater or filament. What would be the effect on the characteristic of reducing the heater current?

3. How many a diode valve be used to obtain a direct voltage from an alternating voltage supply? What is the effect of connecting a reservoir capacitor across the output?

4. Describe the action of a double-diode valve in producing a fully rectified direct voltage supply from an alternating voltage supply. How may the output voltage wave be smoothed?

An introduction to the triode valve

At the same time as Sir Ambrose Fleming was developing the diode valve Dr. Lee de Forest was carrying out experiments in the United States of America. In 1907 he developed the triode valve (Fig. 8.9) which was similar to the diode but included a third electrode in the form of a wire grid mounted between the anode and the cathode. He discovered that when a potential was applied to the grid the flow of electrons through the valve could be regulated.

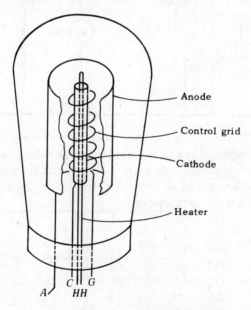

Fig.8.9. Triode valve.

It has already been seen how the diode when used as a rectifier permits current to flow in only one direction through a circuit. The triode valve enables us to control not only the direction but also the magnitude of the current in the anode circuit. A small change in the potential applied to the grid can have quite an appreciable effect on the anode current. This property of the triode may be likened to the small change in pressure required on the accelerator pedal of a motor vehicle to produce a large change in the output developed by the engine.

Triode valve characteristics

Here is an experiment to investigate the effect on the anode current of changes in the anode and grid voltages.

Experiment 8(iv) *To determine the operating characteristic for a triode valve*

In this experiment a circuit was connected as in Fig.8.10.

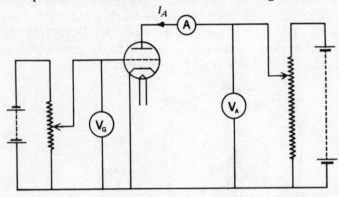

Fig.8.10. Circuit used to determine triode valve characteristics.

With zero voltage applied to the grid the anode voltage was increased in steps and the resulting anode current observed at each stage. This was repeated with voltages of -1 V, -2 V, and -3 V applied to the grid. The observations were as tabulated.

V_G	(V)	0	0	0	0		
V_A	(V)	5	10	15	20		
I_A	(mA)	1·0	2·7	4·8	7·4		
V_G	(V)	−1	−1	−1	−1	−1	−1
V_A	(V)	10	15	20	25	30	35
I_A	(mA)	0·2	0·8	2·0	3·4	5·4	7·5
V_G	(V)	−2	−2	−2	−2	−2	−2
V_A	(V)	20	25	30	35	40	45
I_A	(mA)	0·2	0·7	1·5	2·5	3·9	5·8
V_G	(V)	−3	−3	−3	−3	−3	−3
V_A	(V)	35	40	45	50	55	60
I_A	(mA)	0·6	1·2	2·0	3·3	4·7	6·5

These results are expressed in the graphs shown in Fig.8.11.

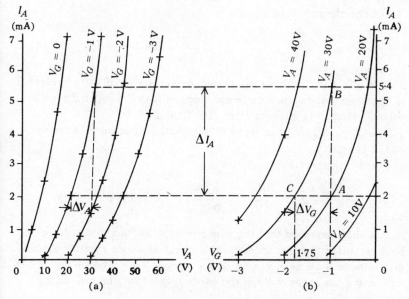

Fig.8.11. Triode valve characteristics.

It is observed that as the grid is made more negative a higher anode voltage is required to maintain the same anode current. The grid is nearer to the cathode than the anode, and when the grid has only a small negative potential applied it has a powerful effect in repelling electrons and reducing the space current.

In Fig.8.11(a) when V_G is -1 V a change in V_A of from 30 V to 20 V produces a change in current I_A of from 5·4 mA to 2·0 mA. This same change in current could be brought about by maintaining V_A constant at 30 V and changing V_G from -1 V to $-1·75$ V (Fig.8.11(b)).

Thermionic valve parameters

In theoretical work involving calculations concerning thermionic valves we find it convenient to refer to certain ratios which may be determined from the valve characteristics. We generally arrange to operate only on the straight parts of the characteristics.

In Fig.8.11(a) it is seen that a change in applied anode voltage ΔV_A will be accompanied by a change in anode current ΔV_A when the voltage applied to the grid remains constant — in this case at -1 volt.

The slope $\Delta V_A / \Delta I_A$ is termed the anode a.c. resistance or anode slope resistance for the valve and is denoted by r_a

For the characteristics plotted,

$$r_a = \frac{\Delta V_A}{\Delta I_A} = \frac{10\,\text{V}}{3\cdot 4\,\text{mA}} = 2940\,\Omega$$

In Fig.8.11(b), with a constant anode voltage of 30 volts applied it is observed that the anode current changes from 5·4 mA to 2 mA when the grid voltage is changed from −1·0 V to −1·75 V.

The slope $\Delta I_A/\Delta V_G$ is termed the mutual conductance of the valve and is denoted by g_m.

For the characteristics plotted,

$$g_m = \frac{\Delta I_A}{\Delta V_G} = \frac{3\cdot 5\,\text{mA}}{0\cdot 75\,\text{V}} = 4\cdot 68\,\text{mA/V}$$

It will be observed that the anode voltage and the grid voltage could both be varied together to maintain a constant anode current. As the grid is made more negative the anode current tends to decrease but by increasing the anode voltage the anode current may be maintained constant.

The amplification factor (μ) of the valve is the ratio of the change in anode voltage to the change in grid voltage which leaves the anode current unchanged. From Fig.8.11(b) it is seen (point A) that the anode current is 2 mA when the grid voltage is −1 volt and the applied anode voltage is 20 volts. If the anode voltage is increased to 30 volts the anode current rises to 5·4 mA (point B). The anode current may now however be reduced to 2·0 mA by adjusting the grid voltage to −1·75 volts The working point is then at C.

$$\mu = \frac{\Delta V_A}{\Delta V_G} \quad \text{which leaves } I_A \text{ unchanged.}$$

$$= \frac{10\,\text{V}}{0\cdot 75\,\text{V}} = 13\cdot 3$$

Note

$$r_a = \frac{\Delta V_A}{\Delta I_A} \quad \text{and} \quad g_m = \frac{\Delta I_A}{\Delta V_G}$$

$$\therefore r_a \times g_m = \frac{\Delta V_A}{\Delta V_G} = \mu$$

Thus the amplification factor is equal to the product of the anode a.c. resistance and the mutual conductance.

From the values of r_a and g_m already obtained for this valve,

$$\mu = 2940 \times 4{\cdot}68/1000 = 13{\cdot}7$$

The difference between this value and the value of 13·3 as obtained earlier is caused by the curvature of the characteristics.

How a triode valve may be used as an amplifier

One application of the triode valve is its use as an amplifier

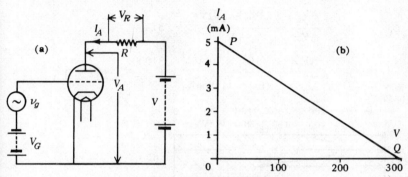

Fig.8.12. Using a triode as an amplifier.

If an alternating voltage v_g is introduced into the grid circuit of a triode valve then the effect is to produce variations in the anode current. When a load resistor R is included in the anode circuit the voltage drop across this resistor varies as the anode current changes. If the magnitude of the change in the voltage across R is greater than the change in grid voltage we have voltage amplification.

Now, it is clear that for a given applied voltage V in the anode circuit the potential between the anode and cathode of the valve will depend on the voltage drop across R and hence on the current in the anode circuit.

$$V = V_A + V_R$$
$$= V_A + I_A R$$

Let us suppose that a d.c. supply of steady value 300 volts is available and that the load resistance in the anode circuit of a valve is $60\,\text{k}\Omega$.

If the grid voltage V_G is made sufficiently negative to reduce the anode current to zero $I_A \times R = 0$ and $V_A = V = 300\,\text{V}$. This enables us to obtain point Q on the diagram in Fig.8.12(b).

When the current is such that the whole of the supply voltage V is dropped across R the voltage V_A will be zero. This enables us to obtain point P in the diagram.

The line PQ is termed the 'load line' for the anode circuit. It shows the relation between the steady anode current and the p.d. across the valve for a given load resistance in the anode circuit.

In Fig.8.13(a) we show typical characteristics relating to a triode valve. Superimposed on these characteristics is the load line PQ for an anode load resistor of 60 kΩ and an anode supply of 300 volts.

$$\mu = \frac{\Delta V_A}{\Delta V_G} = \frac{(107 - 70)\text{V}}{1\text{V}} = 37$$

$$r_a = \frac{\Delta V_A}{\Delta I_A} = \frac{(107 - 70)\text{V}}{(6\cdot 4 - 4\cdot 0)\text{mA}} = \frac{37000}{2\cdot 4} = 15\,420\,\Omega$$

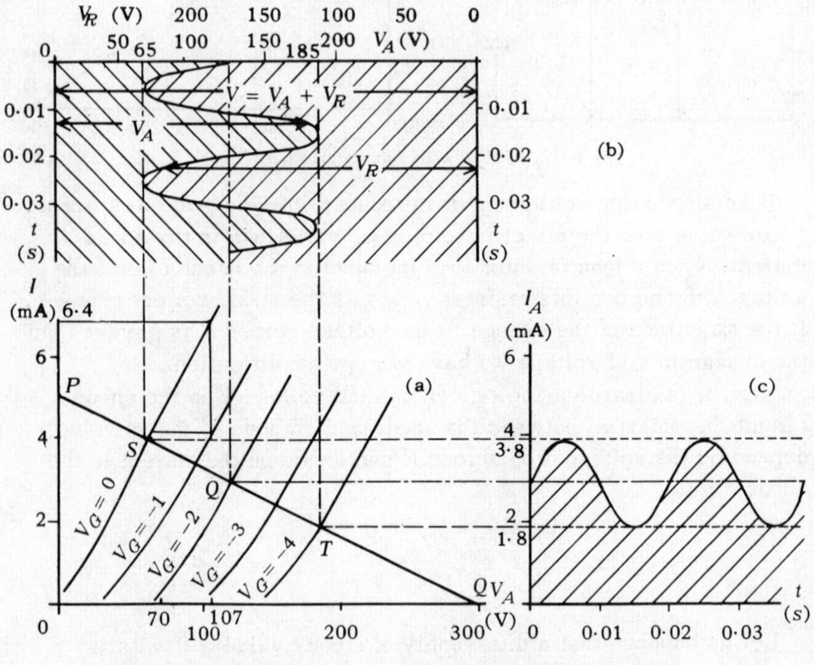

Fig.8.13. Load line construction for an amplifier circuit.

Let us consider the conditions when the grid circuit (Fig.8.12(a)) includes a battery grid bias voltage (V_G) of -2 volts in series with an alternating signal voltage (ν_g) of maximum value 2 volts and frequency 50 Hz. The potential on the control grid of the valve will therefore vary between 0 and −4 volts as shown in Fig.8.14.

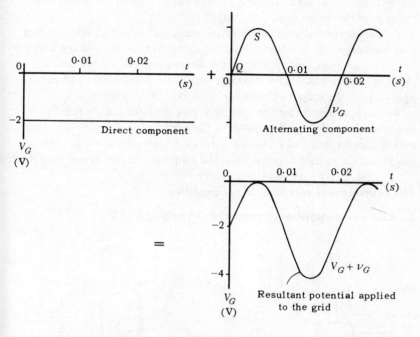

Fig.8.14. Voltages in the grid circuit.

The valve will operate over a range common to both the load line and valve characteristics. As the voltage on the grid varies between 0 and −4 volts the corresponding voltage across the anode and cathode of the valve will vary over a range represented by ST on the load line diagram. (Fig.8.13(a)).

In Fig.8.13(b) we show the variation in the anode voltage with respect to the cathode. The range is from about 65 volts to 185 volts, that is a total swing of 120 volts.

Corresponding to these changes in the anode voltage there are variations in the anode current which are also represented by ST. The maximum and minimum values are approximately 3·8 mA and 1·8 mA respectively (Fig.8.13(c)).

$$\text{The voltage amplification} = \frac{\text{change in output voltage}}{\text{change in input voltage}}$$
$$= \frac{120\,\text{V}}{4\,\text{V}} = 30$$

An important point to note is that as the alternating input to the grid is increasing from Q to S in Fig.8.14. so the operating point on the load line moves from Q to S and the voltage across R, namely V_R is increasing. At the same time the voltage (V_A) between the anode and cathode of the valve is decreasing.

Providing the valve characteristics are parallel and equally spaced over the range ST then the alternating component of the voltage across R will be an amplified version of the alternating voltage introduced into the grid circuit. If the characteristics are not parallel and equally spaced then the output voltage across R will be distorted.

We have assumed that the load is a pure resistor, and hence the amplification will be independent of frequency changes. An inductive coil or tuned circuit may also be used as a load impedance but the amplification is then dependent on the frequency of the signal amplified.

The equivalent circuit for a triode amplifier

A triode valve amplifier circuit is shown in Fig.8.15(a).

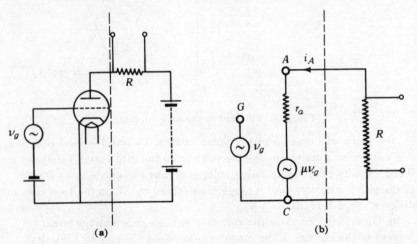

Fig.8.15. Triode amplifier and equivalent circuit.

So far as alternating quantities are concerned we may draw the equivalent circuit as in Fig.8.15(b). The d.c. supplies are replaced by simple conducting paths. It is assumed that the grid voltage is always negative with respect to the cathode and hence with no current in the grid circuit this circuit may be left open at G.

Since $\mu = \Delta V_A / \Delta V_G$ which leaves I_A unchanged, the valve may be looked upon as a generator of output μv_g and internal resistance r_a ohms. Thus the equivalent circuit to the triode is represented by the network with terminals G, A and C. R is a load resistor connected externally to the equivalent circuit.

$$i_A = \frac{\mu v_g}{r_a + R}$$

Instead of using instantaneous values of alternating quantities r.m.s. values may be used,

$$I_A = \frac{\mu V_G}{r_a + R}$$

Let us consider the voltage amplification of the amplifier from the equivalent circuit,

$$\text{voltage amplification (or gain)} = \frac{\text{alternating output voltage}}{\text{alternating input voltage}}$$

$$= \frac{I_A R}{V_G}$$

$$= \frac{\mu V_G R}{(R + r_a) V_G} = \frac{\mu R}{R + r_a}$$

Referring to the previous example where $\mu = 37$, $R = 60\,\text{k}\Omega$ and $r_a = 15\,420\,\Omega$ it is evident that the calculated voltage amplification is given by,

$$\frac{\mu R}{R + r_a} = \frac{37 \times 60}{75 \cdot 42} = 29 \cdot 6$$

This agrees closely with the result obtained from the load line construction.

Equivalent circuits can be extremely useful for simplifying circuits which at first sight appear to be quite involved. As an example let us draw a simplified circuit equivalent to the circuit shown in Fig.8.16(a). It may be assumed that so far as alternating currents are concerned L offers high impedance; the impedance of C is low with respect to R and the impedance of C_2 is small with respect to r_a.
The circuit is simplified in steps as follows.

1. Substitute the A, G, C equivalent circuit for the triode valve (Fig.8.16(b)).
2. Replace any purely d.c. e.m.f's such as the d.c. anode supply voltage, by conducting paths.

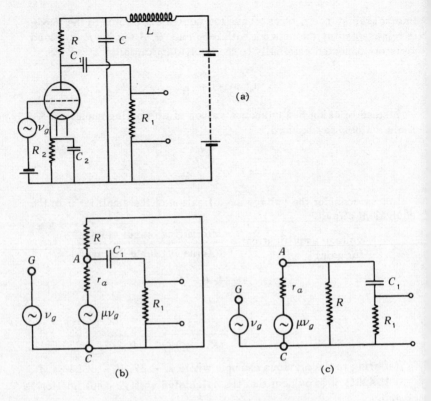

Fig.8.16. Drawing an equivalent circuit.

3. Open circuit any branches in which it is known that the impedance to alternating currents is very high. This includes L. When L is open circuited R and C are in series and since we know the impedance of C is small with respect to R, C may be short circuited.

Since C_2 is small with respect to r_a then C_2 may be short circuited which in turn reduces R_2 to zero. We thus have a simplified equivalent circuit as in Fig.8.16(c).

Multi-stage amplifiers

In an amplifier it is unlikely that the full degree of amplification will be obtained using only one valve — or as we sometimes say in one stage. Usually two or more stages are required and the problem to be faced is shown in Fig.8.17.

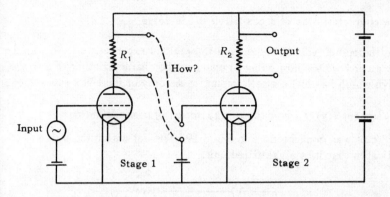

Fig.8.17. Problem of coupling between two stages of an amplifier.

It is essential that only the alternating voltage component of the output from the first stage is fed to the grid circuit of the second stage, since any d.c. component would affect the grid voltage of the second stage. The output across R_1 includes a d.c. component of the anode voltage and also an a.c. component. These components may be separated by using a circuit as in Fig.8.18.

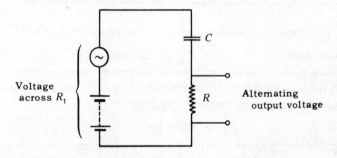

Fig.8.18. Separating direct and alternating voltage components.

If we choose R and C so that R is large with respect to $1/\omega C$ for all frequencies to be dealt with then the alternating component will pass freely through C but not through R and will thus appear across the terminals of R. No direct current can flow through the circuit because of C and hence there will be no direct voltage component across R.

This method of coupling uses an R–C circuit. Alternative systems which you may encounter include R–L and transformer coupling methods.

The characteristics of a gas-filled triode valve

The thermionic valves considered in previous experiments were vacuum valves and we are now going to examine the characteristics of a triode valve which has had a small amount of mercury or inert gas introduced.

Experiment 8 (v) *To investigate the action of a gas-filled triode*

A circuit was connected as in Fig.8.19. The dot within the valve indicates that it is a gas-filled unit.

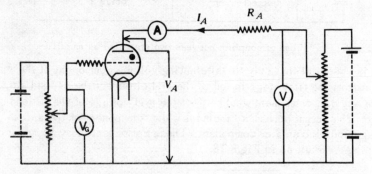

Fig.8.19. Determining the characteristics of a gas-filled triode valve.

With zero voltage applied to the grid, the voltage (V) applied in the anode circuit was gradually increased in steps and corresponding values of the anode current (I_A) observed. The results were as tabulated and are represented by the curve in Fig.8.20.

$$R_A = 1500 \text{ ohms}$$

V	(V)	5	10		15	20	25	30
I_A	(mA)	0·002	0·003	ionisation	2·2	5·7	8·9	12·1
$I_A \times R_A$	(V)	0·003	0·0045		3·3	8·5	13·3	18·2
$V_A = V - I_A R_A$	(V)	4·997	9·9955		11·7	11·5	11·7	11·8

It will be observed that I_A initially increases gradually as V_A is increased. The electrons emitted by the cathode are attracted towards the positive anode. While in transit these electrons colide with atoms of the vapourised gas within the valve. If the electrons move with a sufficiently high velocity then when they collide they cause electrons which are loosely held in the outer orbits of the gas atoms to become detached. This results in more free electrons becoming available within

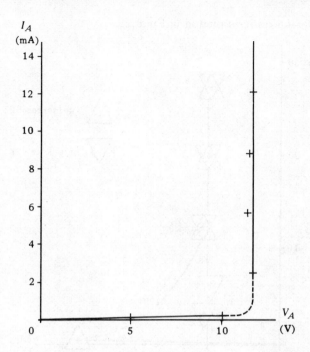

Fig.8.20. Characteristic curve for a gas filled triode.

the valve and as these are attracted towards the anode more collisions take place. We say that the gas becomes ionised and there is a glow discharge within the valve when the valve strikes. The valve resistance is now low and to restrict the anode current a current limiting series resistor R_A is included in the anode circuit.

It is seen that after ionisation has taken place the voltage drop across this valve remains constant at about 11·7 volts. As the voltage applied to the anode circuit is further increased the additional applied voltage appears across the resistor R_A.

The experiment was repeated with −1 volt applied to the grid of the valve and the characteristic plotted for increasing and decreasing values of voltage applied to the anode circuit. Circuit values were as tabulated.

$R_A = 1500$ ohms

V_G	(V)	−1	−1	−1	−1	−1	−1	−1	−1	−1
V	(V)	below 28	28	30	40	50	40	30	20	15
I_A	(mA)	trace	ionisation	13	20	27	20	13	7	3
$I_A \times R_A$	(V)	−		19·5	30	40·5	30	19·5	10·5	4·5
$V_A = V - I_A R_A$	(V)	−		10·5	10	9·5	10	10·5	9·5	10·5

315

The characteristic is plotted in Fig.8.21.

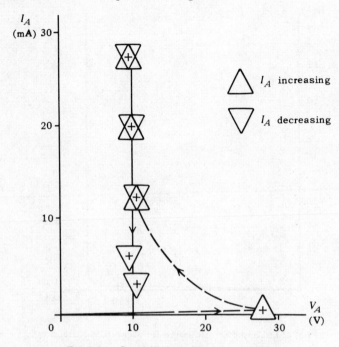

Fig.8.21. Gas filled triode characteristic.

It will be observed from the test results that the voltage at which the valve strikes is higher when the grid carries a negative potential than when the potential on the grid is zero. Once the valve has struck however and the gas become ionised, the voltage may be reduced below that required to produce the initial striking and the gas still remains ionised. The grid is only effective below the striking voltage before the gas has become ionised.

In a further test the voltage V_S required to produce ionisation, with various negative potentials applied to the grid, was investigated. The results were as tabulated and are expressed by the curve of Fig.8.22.

V_G (-V)	0	0·5	1·0	1·5	2·0
V_S (V)	12	16	28	48	88

The mean slope of this characteristic is termed the control ratio.

$$\text{Contol ratio} = \frac{\text{change in anode breakdown voltage}}{\text{corresponding change in grid voltage.}}$$

For the valve tested,
$$\text{control ratio} = \frac{100\,\text{V}}{1\cdot8\,\text{V}} = 55\cdot5$$

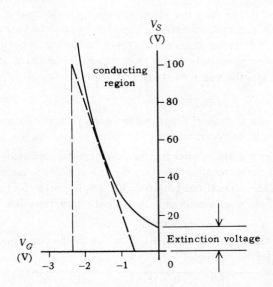

Fig.8.22. Gas filled triode control characteristic.

The gas filled triode has many applications. One of these is its use in time-base circuits and a simple circuit of this type is dealt with in a later section.

Summary

The triode valve includes a third electrode or grid between the anode and the cathode. By varying the potential applied to the grid the anode current through the valve may be varied.

The valve parameters may be expressed as,

$$r_a = \frac{\Delta V_A}{\Delta I_A} \quad g_m = \frac{\Delta I_A}{\Delta V_G} \quad \text{and} \quad \mu = \frac{\Delta V_A}{\Delta V_G}$$

$$\mu = r_a \times g_m$$

A small alternating voltage introduced into the grid circuit can produce a much larger alternating voltage across a load resistor in the anode circuit.

A triode valve may be represented by an equivalent circuit which comprises, so far as alternating quantities are concerned, a generator of output μv_g and internal resistance r_a ohms.

Two or more triodes may be coupled to produce a multi-stage amplifier.

A gas filled triode does not allow any appreciable current to flow until the anode potential is sufficiently high to produce ionisation of the gas. Once the valve strikes however it continues to conduct even though the anode potential may be reduced below the value initially required to cause the valve to strike.

Examples 8.2.

1. Describe with the aid of diagrams the construction and action of a triode thermionic valve. How does the valve differ from the diode?

2. Give a circuit diagram showing the connections necessary to enable us to plot the characteristics for a triode valve. What is meant by anode slope resistance, mutual conductance and amplification factor? Illustrate these terms clearly with reference to typical characteristics.

3. Show how $\mu = r_a \times g_m$ for a triode valve.

Determine the parameters of a triode valve which has characteristics given by the following table:

V_A	50	100	150	200	250	300
V_G	0	0	0			
I_A	1·0	2·8	4·8			
V_G	−1	−1	−1	−1		
I_A	0·3	1·1	2·8	4·5		
V_G		−2	−2	−2	−2	
I_A		0·2	1	2·8	4·8	
V_G			−3	−3	−3	−3
I_A			0·2	1	2·8	4·8

4. How many a triode valve be used as an amplifier? What is meant by the load line for a triode valve? What would be the slope of a load line for a triode valve having an anode load of 80 000 ohms?

5. Explain clearly with the aid of diagrams how the output voltage across an anode load resistor of 75 000 ohms would vary when a sinusoidal alternating voltage of peak value one volt is introduced into the grid circuit of a triode valve. The d.c. grid voltage is −2 volts and the total voltage available in the anode circuit is 300 volts. The valve has characteristics as in Example 3 above.

6. Derive an equivalent a.c. circuit for a triode valve and show how the voltage amplification or gain of a triode amplifier circuit is given by,

$$\frac{\mu R}{R + r_a}$$

7. What problem is involved in coupling the output of one stage in a multi-stage amplifier to the input of the following stage? Sketch a circuit showing how coupling may be achieved.

8. How does a gas filled triode differ from a vacuum triode? Compare the operating characteristics. What is meant by the grid control ratio and the extinction voltage of a gas filled triode?

9. The voltage drop across a gas filled diode remains reasonably constant at 17 volts while conducting over a wide current range. If the valve is connected in series with a 1000 ohm resistor across (a) a 240 V d.c. supply and (b) a 240 V a.c. supply, determine the average current flowing in each case.

Introducing the cathode ray oscilloscope

Now that you have been introduced to diode and triode thermionic valves you are ready to look at the fundamental principle of the cathode ray oscilloscope. This unit has important applications in electrical measurments work and in television receivers.

In the following description of the oscilloscope tube and the circuits associated with it many refinements encountered in a commercial unit have been omitted since our aim, at this stage, is to obtain a simplified outline of how the unit operates.

Experiment 8. (vi) *To investigate the fundamental principle of a cathode ray oscilloscope unit*

The diagram (Fig. 8.23) shows a sketch of an oscilloscope cathode ray tube and also the basic connections which were made to a power supply. It was observed that when the potentiometer B was adjusted a fluorescent spot became visible on the screen and as the potential on the grid (G) was made more negative with respect to the cathode (C) the intensity of the electron beam was reduced. By varying the setting of the potentiometer F the spot could be focussed.

Here is a brief description of how the unit functions.

The cathode is heated by a current passing through the filament and it emits electrons. The electrons are attracted towards the anode A_2 which may commonly have 2–4 kV applied but in the particular unit examined the p.d. was only 800 V. The main purpose of this electrode is to accelerate the electrons. Some of the electrons pass through a hole in the centre of A_2 to strike the fluorescent screen and produce a visible spot. The function of the anode A_1 is to serve as an electron lens and focus the electrons into a narrow beam. By varying the focus

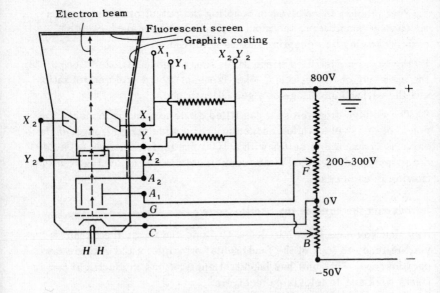

Fig.8.23. Cathode ray tube connections.

control F, the potential of anode A_1 and hence the convergence of the electron beam is varied.

Between the cathode and the first anode is the grid. As in a triode valve a negative voltage applied to the grid controls the number of electrons passing from the cathode to the anode. By adjusting the brilliance or brightness control knob (B) we vary the negative potential applied to the grid, and hence the brightness of the spot on the screen. The grid potential must always be negative with respect to that of the cathode.

Before reaching the fluorescent screen the electron beam passes between two pairs of plates which are set at right angles to each other. The beam may be deflected horizontally by applying a voltage across the X plates and vertically by applying a voltage across the Y plates. The spot may thus be brought to any position on the screen by a suitable combination of steady voltages applied to the deflecting plates. To prevent the screen acquiring a negative charge and repelling further electrons a conducting graphite coating is provided on the inner surface of the tube between the screen and the final anode. The screen is therefore at the same potential as A_2 and electrons do not acquire further acceleration after passing this electrode. The lower part of the tube is surrounded by a mu-metal screen to protect the unit from the effects of external fields.

If you examine the controls of a laboratory C.R.O. unit you ought now to be able to identify the focus and brilliance controls, and appreciate how they act on the beam.

How the position of the spot on a cathode ray tube may be varied

Here is an experiment to measure the sensitivity of a cathode ray tube deflecting system.

Experiment 8 (vii) *To observe the effects of applying external voltages across the plates of a C.R.O. unit*

In this experiment various potentials were applied across the plates of a C.R.O. tube (Fig.8.24) and the effects on the spot observed.

When a p.d. is applied across two plates such as Y_1 and Y_2 electrons passing in a beam between these plates are deflected by the electrostatic force. The electrons are attracted towards the positive plate and repelled from the negative plate.

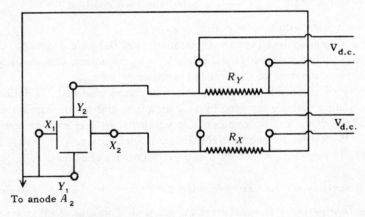

Fig.8.24. Measuring the sensitivity of a cathode ray tube.

Direct potential across R_x

With a p.d. of 30 volts applied from a d.c. source across R_x the horizontal deflection of the spot on the screen was 5 mm. This corresponds to a sensitivity of 0·17 mm/V. It was observed that the direction of the deflection reversed when the connections across R_x were reversed.

Direct potential across R_y

With a d.c. potential of 30 volts applied across R_y the deflection of the spot was observed to be 8 mm. This corresponds to a sensitivity of 0·27 mm/V. The deflection produced when the same voltage is applied to the Y plates is greater than when it is applied to the X plates. This is because the Y plates are further from the screen and also because of differences in the dimensions of the two pairs of plates.

Alternating potential applied across R_y

In this part of the test an alternating voltage of 30 volts was applied across R_y. The spot on the cathode ray tube was observed to trace a vertical line extending 11·5 mm on either side of the zero position of the spot.

The spot moves above and below the zero position depending on the maximum value of the alternating voltage applied across the Y plates. Since 30 volts was the r.m.s. value of the applied voltage
$V_m = 30 \times \sqrt{2} = 30 \times 1\cdot414 = 42\cdot42$ volts.

With a sensitivity of 0·27 V/mm as determined earlier the spot deflection

$$= 42\cdot4\,\text{V} \times 0\cdot27\,\text{mm/V}$$

$$= 11\cdot5 \text{ mm above and below the zero position.}$$

This agrees with the deflection measured.

In a laboratory test instrument, internal amplifiers are generally provided to amplify input signals by a known amount and this enables visual measurements of very small quantities to be made.

Applying potentials across the plates produces electrostatic deflection. When a cathode ray tube is used as a test instrument this is the usual type of deflection employed. In applications to television however it is common to employ electromagnetic deflection.

Here is an experiment to demonstrate magnetic deflection.

Experiment 8 (viii) *To illustrate magnetic deflection of an electron beam*

In the first part of this experiment the external mu-metal shading screen was removed from the tube and when a magnet was moved horizontally towards the tube it was observed that the spot was deflected vertically (Fig.8.25(a)).

An electron beam is a current and when it passes through a magnetic field it has a force acting on it similar to that which acts on a conductor carrying current in a magnetic field.

In the second part of the experiment two coils were fitted to the tube, and it was observed that when a direct current was passed through the coils to establish a horizontal magnetic field, a steady vertical deflection of the spot was produced on the screen (Fig.8.25(b)).

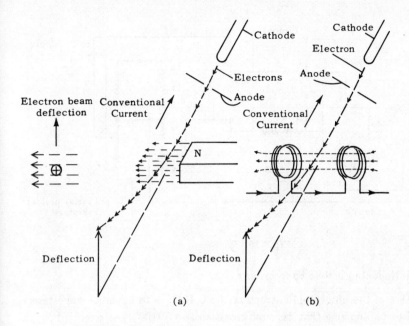

Fig.8.25. Magnetic deflection of an electron beam.

With an alternating current through the coils the spot traced a vertical line on the screen. When the coils were turned to produce a vertical magnetic field then the spot moved to trace a horizontal line.

We have now seen how the spot produced when the electron beam strikes the screen may be moved along either the X and Y axes by electrostatic or electromagnetic deflection.

Shift control

With the X_1 and Y_1 plates connected to anode A_2 as shown in Fig.8.24 and with zero voltages applied across R_x and R_y we would expect the spot to appear in the centre of the screen.

The circuit shown in Fig.8.26 enables us to apply initial steady potentials to the X_2 and Y_2 plates. These potentials differ from the potentials applied to the X_1 and Y_1 plates. The initial zero position of the spot may thus be set to any point on the screen and subsequent variations in p.d. applied across R_x and R_y will lead to further spot deflections along the X and Y axes.

The variable potentiometers P_x and P_y are commonly termed the 'shift' controls. They may be calibrated in terms of the voltage differences which various settings produce between the X_1 and X_2 plates and also between the Y_1 and Y_2 plates.

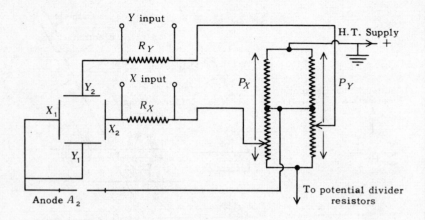

Fig. 8.26. Shift-control circuit.

Introducing a time base circuit

One of the chief applications of the C.R.O. is to examine wave forms. Let us suppose that we wish to examine a 50 Hz waveform.

Consider a voltage with a wave form as in Fig. 8.27 (a) applied across the X plates of a C.R.O. tube. As the voltage increases uniformly from O to P in 1/50 second the spot moves across the screen from A to B (Fig. 8.27 (b)). When the voltage falls from P to Q the spot returns or flies back instantly from B to A.

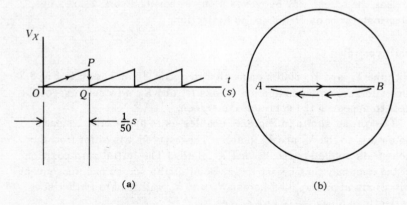

Fig. 8.27. Saw-tooth time base waveform.

Let us assume that the time taken for the spot to move from A to B as the voltage increases form O to P is the time of one cycle of the wave to be examined — in this case 1/50 second.

If the voltage to be examined is applied to the Y plates of the C.R.O. at the same time as the 'saw tooth' voltage OP is applied to the X plates, then the spot will move on the screen from A to B as in Fig.8.28 tracing out the shape of the wave being examined. This process is repeated 50 times per second and the trace appears as a solid line on the screen.

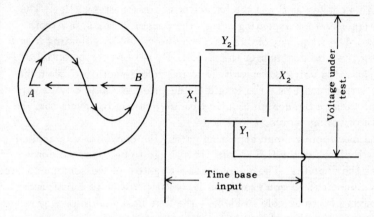

Fig.8.28. Examining an alternating voltage wave.

Our immediate problem is to devise a means of producing a saw tooth type of wave to apply to the X plates as a time base.

The simplest type of circuit which will produce this type of wave is considered in the experiment which follows.

Experiment 8 (ix) *To construct a simple time base circuit for use with a cathode ray tube*

A circuit was connected as in Fig.8.29.

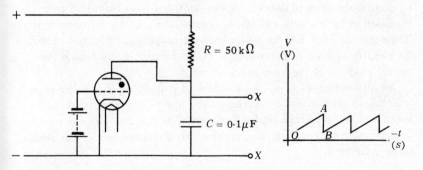

Fig.8.29. Simple time base circuit.

With the resistor R and capacitor C connected in series a charging current flows and the voltage across C rises as represented by OA. A gas filled triode valve is connected across C. When a sufficiently high anode potential is reached, depending on the voltage applied to the triode grid, the valve strikes and the p.d. across the anode and cathode terminals falls very rapidly as represented by AB. Ionisation within the valve ceases at B and the capacitor again acquires charge. The resulting waveform appearing across the capacitor has a saw-tooth outline. The frequency of this wave may be varied by adjusting R or C to alter the time constant of the circuit. By varying the potential on the grid of the valve the magnitude of the voltage across C when the valve strikes and hence the length of the time base axis may be varied. One problem in the design of a time base circuit is to ensure that the growth of voltage across C is linear.

The saw-tooth voltage appearing across the capacitor was applied to the X plates of the C.R.O. and the voltage to be examined connected across the Y plates. The waveform investigated on the screen appeared to be sinusoidal. If you examine a laboratory C.R.O. test instrument you ought now to be able to identify the time base controls. Any range control switch will probably vary the value of C in the time base circuit and any continuously variable control has the effect of varying R. By injecting part of the voltage wave being examined into the grid circuit of the gas filled triode, automatic synchronisation of the time base frequency may be obtained. The control which enables this to be achieved is marked the 'Sync' control.

Summary

The cathode ray oscilloscope tube includes a heated cathode which emits electrons. These electrons are attracted towards and accelerated by positively charged anodes. Some electrons pass through a hole provided in each anode and strike a screen to produce a fluorescent spot The electron beam may be deflected by arranging for it to pass through an electromagnetic or electrostatic field. Shift controls enable the initial position of the spot to be adjusted.

An alternating wave may be examined by applying it to the Y plates which produce vertical deflection of the spot, and by applying a saw-tooth time base voltage to the X plates. The frequency of the time base voltage must be equal to, or a multiple of, the frequency of the voltage to be examined.

Examples 8.3.

1. Sketch a cathode ray oscilloscope tube showing how the electrodes are connected to a potential divider across a d.c. voltage supply. Explain carefully the action of the tube and the focus and brilliance controls.

2. Describe two methods of deflecting the electron beam in a cathode ray tube. Sketch a circuit showing how the shift-control circuit connections may be made to an oscilloscope with electrostatic deflection.

3. What is the purpose of providing a time base circuit for use with a cathode ray oscilloscope? Describe one method of obtaining a time base voltage. How is the time of this supply varied?

9. Semiconductors and Semiconductor Devices

Introduction

Semiconductors are materials with conducting properties between those of good conductors and good insulators. The conducting properties of copper oxide and selenium were known in 1850 and in the early days of broadcasting 'cats whisker' detectors and other semiconductor devices were used. It was only during World War II however that the development of point contact silicon diode detectors for use in microwave radar led to further investigations of semiconductor materials.

Germanium and silicon are two such materials. When they are pure and at low temperatures they have very poor conducting properties. As the temperature is raised however, and particularly when certain impurities are introduced the conductivities increase. Semiconductor materials are used in devices such as solid state rectifiers and transistors.

How does a semiconductor function?

You have already seen in an earlier chapter how an electric current may be considered as the controlled drift of electrons in a circuit. We are now going to look a little closer into this.

In a material such as silicon or germanium the atoms may be considered as being tightly packed and electrons orbit in shells about the nucleus of each atom. Silicon has electrons orbiting in three shells. The inner shell has two electrons the middle shell has eight electrons and the outer shell four electrons (Fig. 9.1(a)).

In germanium each atom has 32 electrons moving in four shells (Fig. 9.1(b)). As with the silicon atom there are four electrons in the outermost shell. Electrons in this outer shell are known as valence electrons.

Let us consider adjacent atoms of silicon or germanium. Each atom shares its valence electrons with four adjoining atoms (Fig. 9.2). Atom A is shown linked to atom B by the electrons e_A and e_B circulating in a common orbit around A and B. This is known as covalent bonding

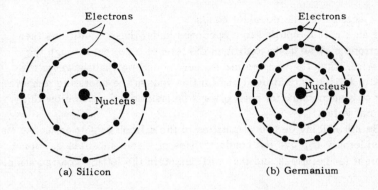

Fig. 9.1. Silicon and germanium atoms.

and every atom is bonded to four neighbours by shared electrons. Thus in addition to being bonded to atom A, atom B is also bonded to the atoms C, D and E. For simplicity this bonding is represented in the diagram by horizontal and vertical connecting lines. When all the atoms in a material are linked in this manner they are said to form a lattice network.

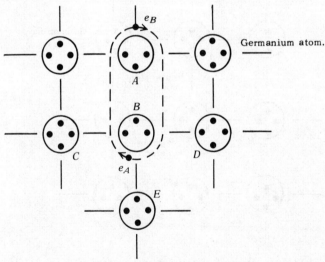

Fig. 9.2. Covalent bonding.

When silicon and germanium are pure these bonds are very strong. There are no free electrons and the materials behave as insulators. Under certain circumstances such as when the silicon or germanium is subjected to heat or illumination, energy may be imparted to the electrons in the outer orbits to such an extent that some of these electrons

329

are ejected from the covalent bonds.

If an e.m.f. is applied to a specimen under these conditions then electrons may be made to drift in the form of a current through the material. When the temperature exceeds a certain limit however the crystal lattice is destroyed and for this reason the operating temperature of a semiconductor must always be maintained well below this critical value.

By introducing certain impurities to the silicon or germanium we may considerably improve the conductivities of these materials at normal ambient temperatures and thus still maintain the lattice configuration.

N-type semiconductors

An n-type semiconductor has impurities such as arsenic or antimony added to the silicon or germanium. These impurities have five valence electrons in the outermost orbit and are said to be pentavalent. Only four of the electrons are required to link into covalent bonds with neighbouring silicon or germanium atoms A, B, C and D (Fig. 9.3).

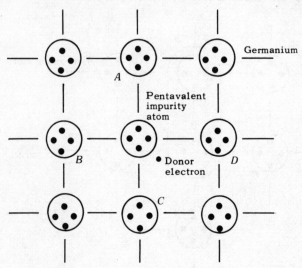

Fig. 9.3. N-type material with pentavalent impurity.

This type of impurity which provides additional electrons is known as a donor and the semiconductor with additional electrons is referred to as n-type material (Fig. 9.4). Each atom of impurity provides one electron which is free to move as a current carrier.

When an e.m.f. is applied across n-type material (Fig. 9.5) a current flows and this may be looked upon as a movement of negative charge carriers or electrons.

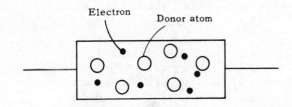

Fig. 9.4. N-type semiconductor.

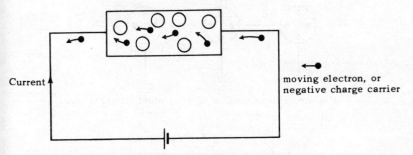

Fig. 9.5. Current flow through n-type semiconductor.

Since the total charge on an atom is zero then when an electron or negative charge carrier is removed the remaining part of the atom must exhibit a positive charge. It attracts any other free electron available.

P-type semiconductors

P-type semiconductor material has impurities such as indium, gallium or boron introduced. These are trivalent elements with three valence electrons in the outer orbit. When they link with neighbouring silicon or germanium atoms the covalent structure is incomplete. Each impurity atom introduces a vacancy for an electron and we say that holes are introduced into the material (Fig. 9.6).

The impurities in the material accept electrons to neutralise the holes and are known as acceptors. This semiconductor with holes for electrons is known as p-type material.

When an e.m.f. is applied across p-type material (Fig. 9.8) current flows and this current may be looked upon as a movement of holes or positive charge carriers through the material.

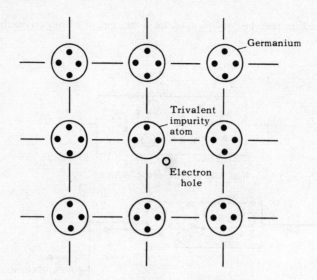

Fig. 9.6. *P*-type material with trivalent impurity.

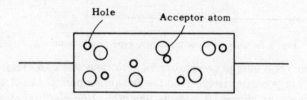

Fig. 9.7. *P*-type semiconductor.

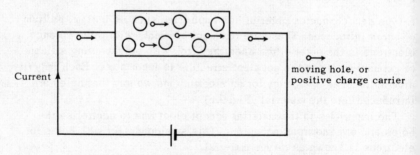

Fig. 9.8. Current flow through *p*-type semiconductor.

Since the total charge on an atom is zero then when a hole or positive charge carrier is removed the remaining part of the atom must exhibit a negative charge and it attracts any other free hole available.

What happens at the junction between p and n-type materials when a voltage is applied?

Let us investigate this experimentally.

Experiment 9 (i) *To investigate the current across the junction between p and n-type semiconductors when a voltage is applied*

The semiconductor materials used in this experiment were arranged as in Fig. 9.9. It will be seen later that this arrangement constitutes a transistor but at this stage we only concern ourselves with the individual junctions between the materials.

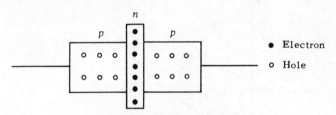

Fig. 9.9. Junction transistor.

A circuit was connected as represented by Fig. 9.10.

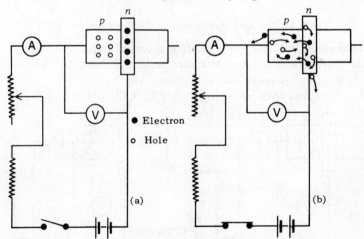

Fig. 9.10. Effect of applying a forward bias potential across a p-n junction.

Diagram (a) of Fig. 9.9 shows the conditions which exist when the switch is open. Both the p and n-type semiconductors have zero total charges but the p-type material has holes which may be looked upon as positive charge carriers and the n-type material has electrons or

333

negative charge carriers. When the switch was closed as in (b) it was observed that current flowed in the circuit. The positive terminal of the applied voltage attracted electrons or negative charge carriers while the negative side attracted the holes or positive charge carriers.

The observations were as tabulated and Fig. 9.11 shows how the current varies with increasing applied p.d. across the junction.

V (V)	0·1	0·2	0·3	0·4	0·5	0·6
I (mA)	0·1	0·3	0·6	0·9	1·5	2·1

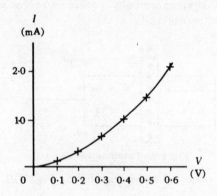

Fig. 9.11. Forward bias characteristic for a *p-n* junction.

This method of applying a positive potential from a d.c. supply to the *p*-type semiconductor is known as forward biasing. The junction offers a fairly low resistance to current flow.

A circuit was next connected as in Fig. 9.12.

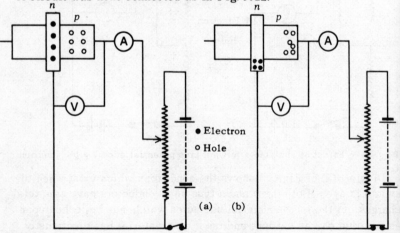

Fig. 9.12. Effect of applying a reverse bias potential across an *n-p* junction.

When the switch is closed the negative charge carriers in the n-type semiconductor are attracted towards the positive side of the applied voltage while the positive charge carriers in the p-type semiconductor material are attracted towards the negative terminal of the applied voltage. The result is that there are no charge carriers at the junction between the n and p-type materials and the junction offers a high resistance to current flow. A voltage applied in this direction across an n-p junction is known as negative biasing.

The results of increasing the applied voltage were as tabulated and the graph (Fig. 9.13) shows the relationship between the applied voltage and the reverse current.

V (V)	–0·5	–1·0	–1·5	–2	–2·5	–3·0	–4·0
I (μA)	2·0	2·5	2·8	3·0	3·1	3·15	3·2

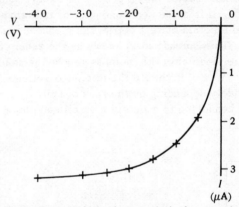

Fig. 9.13. Reverse bias characteristic for a p-n junction.

We thus see how the junction between p and n-type semiconductors offers high resistance to the flow of current in one direction and a comparatively low resistance in the opposite direction. This is a rectifying property and semiconductors may be used as devices which perform similar functions to those of diode thermionic valves.

The reverse current characteristic for a semiconductor junction depends on the temperature at which the semiconductor is operating. For a germanium device the reverse current approximately doubles for every 10°C rise in temperature, while for silicon it doubles for a temperature rise of about 8°C. At a given temperature the reverse current for germanium is much lower than for silicon and the silicon diode can operate at a much higher temperature than germanium. In practice the temperature of a germanium device is restricted to about

55°C – 60°C. The power loss is dissipated by fitting copper or aluminium cooling fins and using natural or forced air cooling.

When considering electrical instruments we saw how the barrier layer between selenium and a metal plate on which it was deposited had rectifying properties. The advantage of using germanium or silicon semiconductors is that much higher current densities may be employed. Germanium may for example be operated at $1\,\text{A}/\text{mm}^2$ while selenium is operated at about $2\cdot5\,\text{mA}/\text{mm}^2$.

Stabilising a voltage supply using a semiconductor diode

One particular semiconductor device is the zener diode. With forward bias applied the device operates with a normal semiconductor junction characteristic. If a reverse voltage is applied and increased, the energy of the current carriers may be brought to a point where additional carriers are dislodged from the semiconductor by collisions. An avalanche of hole and electron current carriers is produced and the device becomes ionised. This breakdown is known as the zener effect. The voltage across the diode after this point is reached remains fairly constant for a given unit and is termed the reference voltage. The reference voltage varies widely depending on the type of unit.

A zener diode connected to maintain a stabilised output voltage is shown in Fig. 9.14.

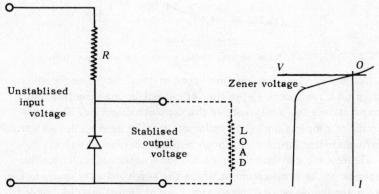

Fig. 9.14. Using a zener diode to maintain a stabilised voltage.

If the supply voltage increases then the current increases, additional voltage is dropped across R, and the diode maintains a fairly constant voltage across the output terminals. Care must be taken to ensure that the maximum permissible temperature rise is not exceeded.

What is a transistor?

It was 40 years after Lee de Forest invented the triode valve that Bardeen and Brattain invented the point contact transistor. Later developments produced the junction transistor.

The advent of the transistor in 1948 saw the beginning of an era in which the rapid growth of transistor electronics was to have repercussions throughout the whole field of electrical engineering. Transistors can perform many of the functions for which thermionic valves were formerly used. They have advantages in that they are smaller than thermionic valves, they do not require heater supplies and operate at much lower voltages. The use of transistors in hearing aids, in pocket and portable type communications equipment, in microphone preamplifiers and in computers are but a few of the everyday applications of these devices.

A transistor is a device by which a current may be transferred from a low resistance circuit to a high resistance circuit with a consequent gain in power. The term 'transistor' is a combination of the words transfer and resistor.

An alloy p-n-p transistor consists of a layer of n-type material sandwiched between two layers of p-type material. In an n-p-n type transistor a layer of p-type material is sandwiched between two layers of n-type material. Figure 9.15 shows a p-n-p transistor.

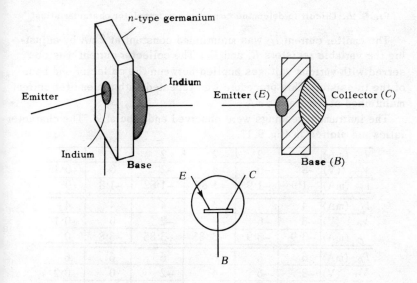

Fig. 9.15. Junction type p-n-p transistor.

The n-type material in the centre is known as the base material and may be about 0·1 mm thick. On either side of this we have the collector and emitter. These may be in the form of indium beads attached to the base during a heat treatment process which produces the p-type material. The collector is larger than the emitter to assist in dissipating the power loss which is greater at the collector-base junction than at the emitter-base junction.

How does a transistor function?

Experiment 9(ii) *To plot the static characteristics for a common-base n-p-n transistor*

A circuit was connected as shown in the diagram (Fig. 9.15).

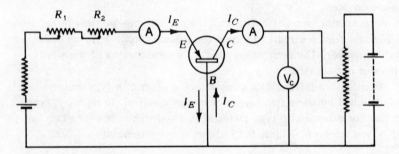

Fig. 9.16. Circuit to determine common base transistor characteristics.

The emitter current I_E was maintained constant at 2 mA by adjusting the variable resistors R_1 and R_2. The collector current was observed with various voltages applied between the collector and base of the transistor. This procedure was repeated with the emitter current maintained at 4 mA and 6 mA.

The instrument readings were observed and tabulated. The characteristics are plotted in Fig. 9.17.

I_E (mA)	2	2	2	2	2	2
V_C (V)	−8	−6	−4	−2	0	+0·15
I_C (mA)	−1·9	−1·9	−1·85	−1·82	−1·8	0
I_E (mA)	4	4	4	4	4	4
V_C (V)	−8	−6	−4	−2	0	+0·18
I_C (mA)	−3·9	−3·9	−3·85	−3·85	−3·8	0
I_E (mA)	6	6	6	6	6	6
V_C (V)	−8	−6	−4	−2	0	+0·2
I (mA)	5·8	5·8	5·75	5·75	5·7	0

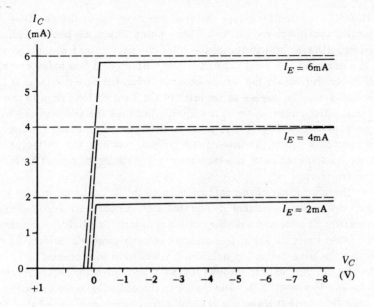

Fig. 9.17. Common base transistor characteristics.

Let us see if we can account for the shape of these characteristics.

Consider the simplified diagram shown in Fig. 9.18 where the emitter-base junction is biased in the forward direction and the collector-base junction is biased in the reverse direction.

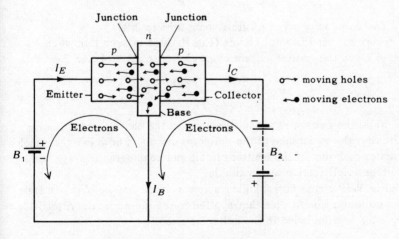

Fig. 9.18. Simplified transistor circuit.

Holes in the emitter p-type material are repelled by the positive potential applied to the emitter. These holes which are positive charge carriers, diffuse through the base-collector junction and are attracted by the negative potential applied to the collector. A few holes combine with the electrons in the n-type base material and $I_E = I_C + I_B$. In practice since the degree of impurity in the base is less than in the collector and emitter elements, and also because the width of the base is kept as narrow as possible there is little recombination of holes and electrons within the base. Each hole on reaching the collector p-type material requires an electron from the collector circuit battery. to neutralise it.

As the magnitude of the applied voltage in the collector circuit is varied the collector current I_C remains almost constant. Because of the emitter to base current, the collector current is slightly less than the emitter current. When the collector voltage is zero a current flows through the base-collector junction. This current is produced by holes from the emitter passing through the base and collector before returning to B_1. If a small p.d. is included in the collector base circuit to oppose this current flow then the collector current may be reduced to zero.

Any change in current (ΔI_E) in the emitter circuit will be accompanied by a change in current (ΔI_C) in the collector circuit. The ratio of these changes for a constant collector voltage is termed the current gain or current amplification and denoted by

$$\alpha = \frac{\Delta I_C}{\Delta I_E}$$

The value of α may vary from about 0·85 to 0·99.

From our experimental curves (Fig. 9.16) it is seen that when V_C is −4 V and the emitter current changes from 6 mA to 4 mA,

$$\alpha = \frac{(5\cdot75 - 3\cdot85)}{2\,\text{mA}} = 0\cdot95$$

While the current is almost the same in the emitter and collector circuits, the resistance of the collector circuit is however very much greater than that of the emitter circuit and considerable power and voltage amplification are available.

Fig. 9.19 shows how a higher output voltage may be obtained from the collector circuit than that applied to the input circuit. A load resistor R is included in the collector circuit.

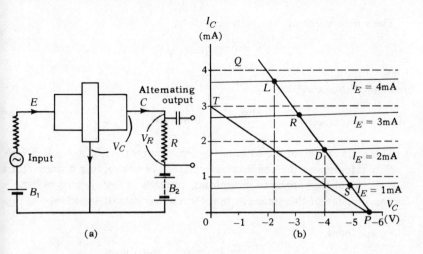

Fig. 9.19. Transistor common base amplifier circuit and load line diagram.

Let us consider as an example a family of transistor characteristics as in Fig. 9.19(b). If the potential on the collector is −4 volts when the input signal to the emitter is zero and the emitter current is 2 mA the conditions will be as represented by point D. The collector current I_C is 1·8 mA. If the resistance of the load is 1000 ohms the p.d. across R will be given by $1000\,\Omega \times 1\cdot 8\,\text{mA} = 1\cdot 8\,\text{V}$.

The total bias which must be provided by B_2 is $-4 -1\cdot 8 = 5\cdot 8\,\text{V}$. This is represented by point P.

A line through P and D to Q enables us to determine the collector current for different values of the emitter current providing that the collector-circuit voltage is maintained constant. $P-Q$ is termed a load line and has a slop of $1/R$ where R is the load resistance.

For example, if the emitter current is increased to 4 mA the collector current which is denoted by L is 3·6 mA, V_C is −2·2 V and V_R is 3·6 V.

Let us introduce an alternating current of maximum value 1 mA into the emitter circuit. The collector current varies between points R and S on the load line. The alternating collector current through R varies between 0·85 mA and 2·7 mA.

The peak value of the alternating voltage component across R will be given by

$$\frac{(\text{change in } I_C)}{2} \times R = \frac{(2\cdot 7 - 0\cdot 85)\,\text{mA}}{2} \times 1000\,\Omega$$

$$= 925\,\text{mV} = 0\cdot 025\,\text{V}$$

The r.m.s. value of this voltage will be

$$\frac{0\cdot 925}{\sqrt{2}} = 0\cdot 654 \text{ V}$$

and hence the power in $R = V^2/R$

$$= \frac{0\cdot 654^2}{1000} = 0\cdot 425 \text{ mW}$$

The r.m.s. value of the alternating input current $= 1/\sqrt{2}\text{mA} = 0\cdot 707\text{mA}$

Now a feature of the common base method of connecting a transistor is that the input impedance to alternating currents is low. Let us assume for the purpose of this example that the emitter circuit impedance is 25 ohms.

Power input,

$$I_E^2 R_E = \left(\frac{0\cdot 707}{1000}\right)^2 \times 25 = 0\cdot 0125 \text{mW}$$

This is a power gain of

$$\frac{0\cdot 425 \text{ W}}{0\cdot 0125 \text{ W}} = 34$$

What about the voltage gain?

Input, $V_E = I_E \times R_E = 0\cdot 707 \times 25 = 17\cdot 65 \text{ mV}$
Output, $V_C = I_C \times R_C = 0\cdot 654 \text{ V}$

$$\text{Voltage gain} = \frac{0\cdot 654}{17\cdot 65} \times 1000 = 37$$

The higher we make R_L for a given value of V_C the higher the output voltage and the output power. The maximum value of R_L which can be used is limited however by the maximum permissible distortion.

With a p.d. of 5·8 volts applied in the collector circuit and a load resistor of 2000 ohms the load line would be represented by PT which has a slope of 1/2000 or 2·9 mA/5·8 V.

Assuming the same emitter circuit resistance, students should show that the power and voltage gains are now approximately 64 and 72 respectively.

Common base connections of a transistor provide a current gain of less than unity, a low input impedance and a high output impedance, There is a small collector leakage current.

Connecting a transistor with common emitter circuit connections

In the previous experiment we considered a transistor with the base common to the emitter and collector circuits. This is known as the common base circuit connection and the current gain is less than unity. We now come to consider an alternative connection where the emitter is common to the input and output circuits. This device may be used as a current amplifier.

Experiment 9 (iii) *To determine the static characteristics for a common emitter transistor circuit*

A circuit was connected as in Fig. 9.20.

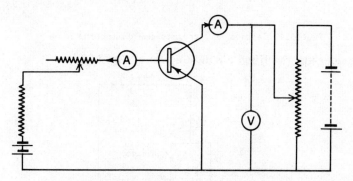

Fig. 9.20. Measuring common emitter characteristics.

The base current was maintained constant at (a) zero, (b) $40\mu A$ and (c) $80\mu A$. The collector voltage was increased in steps to 8 volts and the collector currents observed.

The results were as tabulated and are shown plotted in Fig. 9.21(a).

I_B (μA)	0	0	0	0
V_C (V)	−1·0	−2·0	−4	−6
I_C (mA)	0·125	0·13	0·145	0·16
I_B (μA)	40	40	40	40
V_C (V)	−1	−2	−4	−6
I_C (mA)	2·0	2·1	2·25	2·35
I_B (μA)	80	80	80	80
V_C (V)	−1	−2	−4	−6
I (mA)	4·4	4·6	4·9	5·1

How can we account for the working of the transistor as a current amplifier?

343

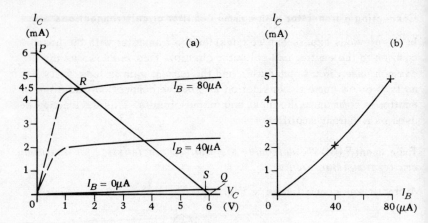

Fig. 9.21. Common emitter transistor characteristics.

Consider the simplified circuit diagram of Fig. 9.22.

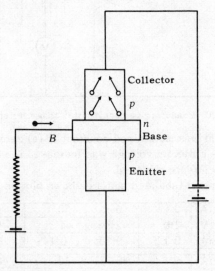

Fig. 9.22. Common emitter circuit.

The input circuit is associated with the injection of current carriers into the base and the output circuit with the flow of carriers from the emitter to the collector through the battery. Electrons flow into the base at B and replace electrons which combine with holes. When the electrons supplied to the base are plentiful the output current increases but if only a few electrons are available the output emitter current is small.

For a constant collector voltage small variations in base current produced comparatively large changes in the collector current. If we plot collector current against base current as in Fig. 9.21(b) we obtain the transfer characteristic and the current amplifications α' or β is given by the ratio,

$$\beta = \frac{\text{change in output current}}{\text{change in input current}}.$$

In this circuit the base current is the controlling element. From the curves, when $V_C = -4$ volts,

$$\beta = \frac{\Delta I_C}{\Delta I_B} = \frac{(4 \cdot 9 - 2 \cdot 25)\,\text{mA}}{(80 - 40)\,\mu\text{A}} = \frac{2 \cdot 65 \times 10^6}{40 \times 10^3} = 66 \cdot 3$$

Let us consider the effect of including a 1000 ohm load resistor in the collector circuit of a transistor with characteristics as in Fig. 9.21. An alternating current of peak value 40μA is to be introduced into the base circuit. The steady base current is 40μA and a p.d. of 6 V is available in the collector circuit.

The load line will be shown by PQ,

$$\frac{1}{R} = \frac{I}{V} \text{ that is } \frac{1}{1000\Omega} = \frac{0 \cdot 006\,\text{A}}{6\,\text{V}} = \frac{6\,\text{mA}}{6\,\text{V}}$$

The collector circuit current will vary between $4 \cdot 5$ mA and $0 \cdot 2$ mA while the base circuit current varies between R and S, that is between 0μA and 80μA. The p.d. across the load resistor will be alternating and of r.m.s. value $(4 \cdot 5 - 0 \cdot 2)\,\text{mA}/2\sqrt{2} \times 1000\Omega = 1 \cdot 52\text{V}$

$$\text{Current amplification} = \frac{4 \cdot 3\,\text{mA}}{80\,\mu\text{A}} = 54$$

Note

We have expressed the power gain of an amplifier circuit as the ratio of the output power to the input power. An alternative method is to express the power change in terms of bels or decibels.

$$\text{Power gain} = \log_{10} \frac{\text{power output}}{\text{power input}} \text{ bels}$$

$$= 10 \log_{10} \frac{\text{power output}}{\text{power input}} \text{ decibels (dB)}$$

Summary

Pure materials such as silicon and germanium have poor conducting properties at low temperatures.

By introducing impurities which are pentavalent, conduction is improved and with donor electrons the material is known as n-type material. Alternatively, impurities which are trivalent may be introduced and these improve conduction by providing holes which readily accept electrons. These impurities are known as acceptors and the semiconductor material is p-type material.

When a voltage is applied across a p - n junction the resistance presented by the junction depends on the polarity of the voltage. When the positive side of the supply is connected to the p-type semiconductor current flows readily, but biasing a junction in the reverse direction produces only a minute current flow.

A semiconductor zener diode with negative biasing may be used as a voltage stabiliser.

A transistor consists of a sandwich of p-n-p or n-p-n materials. With common base connections the current gain is slightly less than unity but high voltage and power gains may be achieved. In the common emitter circuit, high current gain results.

A power gain expressed in decibels is given by $10 \log_{10}$ power output/power input.

Examples 9.1.

1. Explain the meaning of covalent bonding by referring to germanium atoms. How may pure germanium be converted to (a) n-type and (b) p-type semiconductor material?

2. What is meant by (a) forward and (b) reverse biasing of a p-n junction. Explain the difference in the effects and a practical application of this phenomenon.

3. How may a voltage be stabilised using a zener diode?

4. Describe the construction of a p-n-p type transistor. Sketch a circuit suitable for determining the characteristics of a transistor connected with common base connections. What is the magnitude of the current gain with this method of connection? How may voltage or power amplification be achieved?

5. Describe the common emitter method of connecting a transistor. Give a circuit diagram suitable for use in obtaining the common emitter characteristics and sketch typical characteristics. The characteristics of a transistor are represented by the following table:

Collector voltage (V)	Collector current (mA)		
	$I_B = 20\mu A$	$I_B = 40\mu A$	$I_B = 60\mu A$
1	1·2	2·0	2·8
6	1·55	2·35	3·15
10	1·8	2·6	3·4

If the transistor is connected as a common emitter amplifier with a load resistance of 2000 ohms in the collector circuit determine, (a) the total voltage swing across the load for an input signal current of peak value $20\mu A$ (b) the steady collector voltage and (c) the current gain. A supply of 9 volts is available for the load circuit and the base bias current is $40\mu A$.

6. Compare the current gains obtainable with an input signal of peak value $20\mu A$ when the load resistor in the transistor circuit of Example 5 is a (a) 1500 ohms and (b) 3000 ohms. A collector voltage of 9 volts is available. What factors limit the gain obtainable from a transistor amplifier stage?

7. What is meant by the decibel?

Determine the power gain in decibels for an amplifier if a change in the input circuit current of 1 mA produces a change in the output load of 0·9 mA. Input and load circuit resistances are 80 and 2500 ohms respectively. What is the numerical gain corresponding to (a) 1 decibel, (b) 2 decibels and (c) 1 bel?

10. Illumination and Electric Lamp Circuits

Introduction

It is generally accepted that light is a form of energy. If the temperature of a solid body is sufficiently high it becomes 'red-hot' and emits light waves which appear red in colour. Radiated waves from a body may extend over a very wide spectrum and those associated with light as visible to the human eye cover only a very small section of the electromagnetic spectrum.

The diagram (Fig. 10.1) shows in outline the range of waves in the electromagnetic spectrum. At the low frequency end we have the audible frequencies, and then passing through the radio frequency range we come into the infra-red region. Waves in this section are felt as heat by the human body. In the visible spectrum impressions of colour depend on the radiation frequencies. Higher frequencies bring us into the ultra-violet region where these waves produce fluorescence in materials such as are used to coat discharge lamp tubes and television screens. Above that range we pass into the region of X-rays and cosmic rays.

In this chapter we concern ourselves with lamps emitting rays which are in the visible spectrum, extending from about $3 \cdot 9 \times 10^{14}$ Hz to about $7 \cdot 5 \times 10^{14}$ Hz. We will consider various types of electric lamp but must first deal with photometry which will introduce us to the units used in illumination calculations.

How the illumination on a surface depends on the distance of the surface from the light source

Here is an experiment to illustrate this.

Experiment 10 (i) *To show how the illumination at a point on a surface depends on the distance of the light source from the surface*

Apparatus was set up as shown in Fig. 10.2.

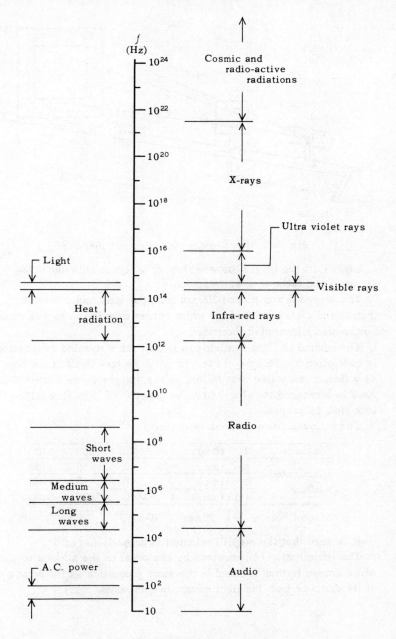

Fig. 10.1. The electromagnetic spectrum.

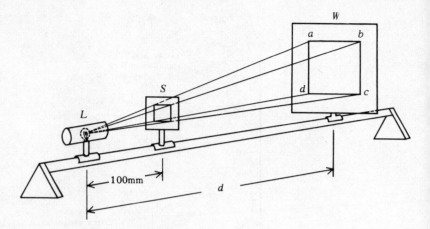

Fig. 10.2. How illumination varies with distance.

Light from the lamp L passes through a square hole cut in the screen S and falls on a white screen W.

The screen S was placed 100 mm from the lamp and it was observed that as the distance (d) of the white screen from the lamp was increased so the area illuminated increased.

The amount of light available to illuminate W remains constant and is controlled by the size of the hole in S. We may think of the light as a flux of many fine rays falling on the surface being illuminated. As d is increased the illumination or quantity of light flux falling on unit area decreases.

The following observations were made:

Distance	d (mm)	200	300	400
Illuminated surface	ab (mm)	40	60	80
	bc (mm)	40	60	80
	area (mm²)	1600	3600	6400

Size of hole in S = 20 mm × 20 mm.

It is seen that the area illuminated is proportional to d^2.

The illumination (E) received by any point on the surface of the white screen is thus reduced in the same proportion as the square of its distance from the light source is increased.

$$E \propto \frac{1}{d^2}$$

Luminous intensity (I)

The *candela* (cd) is the basic SI unit of luminous intensity which is represented by the symbol I. We require a standard by which to measure the unit of luminous intensity.

The candela is the luminous intensity in the perpendicular direction, of a body with a surface area of $1/600\,000\,m^2$ when at the temperature at which platinum changes from the liquid to the solid state under specified pressure conditions.

National laboratories in several countries have set up secondary standards of luminous intensity in the form of specially designed tungsten lamps calibrated against the primary standard. In photometry work light sources are assumed to be point sources although of course this is not always realisable in practice.

The luminous intensity of two lamps in a given direction may be compared by a simple experiment.

Experiment 10 (ii) *To compare the luminous intensity of two lamps*

A simple photometer bench was set up in a darkened room as in Fig. 10.3.

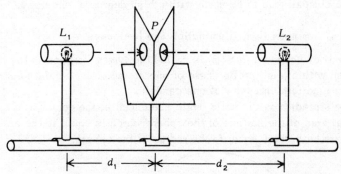

Fig. 10.3. Comparing luminous intensities.

A piece of folded paper was placed on the stand between the lamps L_1 and L_2 which were to be compared. The stand with P was moved until both sides of the paper were illuminated equally when viewed from the front.

Equal illumination was received with $d_1 = 0{\cdot}28\,\text{m}$ and $d_2 = 0{\cdot}72\,\text{m}$.

Now, it will be appreciated that the illumination received by a surface depends upon the luminous intensity I of a light source. The illumination was also seen in the previous experiment to vary inversely with the square of the distance between the source and the illuminated surface.

$$E_1 \propto \frac{I_1}{d_1^2} \quad E_2 \propto \frac{I_2}{d_2^2}$$

When equal illumination is received from each lamp,

$$\frac{I_2}{I_1} = \frac{d_2^2}{d_1^2}$$

$$\frac{I_2}{I_1} = \frac{0 \cdot 72^2}{0 \cdot 28^2} \text{ or } I_2 = 6 \cdot 6 \, I_1$$

If one of the lamps is a standard lamp of known luminous intensity then the luminous intensity of the second lamp may be calculated.

Luminous intensity of L_2 = luminous intensity of $L_1 \times (d_2/d_1)^2$
Using a piece of folded paper as in this experiment to receive illumination from the two sources enables us to obtain only an approximate comparison between the luminous intensities of the two lamps. Various optical arrangements termed photometer heads have been designed to enable comparisons to be made with a high degree of accuracy.

Units of luminous flux, illumination and luminance

The unit of *luminous flux* (ϕ) is called the *lumen* (lm) and is the flux emitted within a unit solid angle of one steredian by a point source having a uniform intensity of one candela.

The steradian is the solid angle subtended at the centre of a sphere by that area of the surface of the sphere which is equal to the area of a square with sides equal to the radius of the sphere.

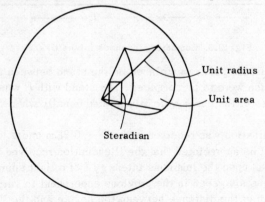

Fig. 10.4. Steradian or unit solid angle.

It is thus evident that the total luminous flux emitted by a point source of one candela is 4π lumens.

The *illumination* (E) at any point on a surface is the luminous flux acting on unit area of the surface at that point. The unit of illumination is the *lux* (lx) which is equal to one lumen per square metre (lm/m^2).

The *luminance* (L) or brightness of a surface which may be either illuminated or self luminous is measured in terms of candelas per square metre (cd/m^2). Sometimes this unit is termed the *nit*.

Approximate values of luminance for various sources are as follows:

Source	Luminance (cd/m^2)
Tungsten clear lamp	550 to 650 $\times$ 10^4
Tungsten pearl lamp	8 to 9 $\times$ 10^4
Mercury low pressure fluorescent	0·5 to 0·9 $\times$ 10^4
Clear blue sky	0·4 $\times$ 10^4

If there is a considerable contrast in the luminance of adjacent objects viewed, one experiences glare and discomfort. From this point of view it will be appreciated that the pearl type of filament lamp and the fluorescent lamp are improvements over the clear type of filament lamp.

The inverse square and cosine laws of illumination

We have already seen that the illumination of a surface,

$$E \propto \frac{I}{d^2}$$

where I is the luminous intensity of the light source and d the distance between the source and surface.

Also, a source of intensity I cd produces an illumination of I lm/m^2 or lx, at a distance of one metre.

$$\therefore E = \frac{I}{d^2} \; lm/m^2 \text{ or lux.}$$

This expression for the illumination received by a surface is known as the inverse square law.

Let us now consider the case where the illuminated surface is tilted through an angle θ to the normal as in Fig. 10.5. The incident light rays are no longer perpendicular to the surface but meet it at the angle θ.

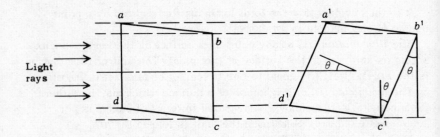

Fig. 10.5. Effect of tilting the illuminated surface.

The same flux now falls on the increased area $a'b'c'd'$ compared with the former area $abcd$, and hence the illumination E in lm/m² is reduced.

$$\frac{\text{surface area } abcd}{\text{surface area } a'b'c'd'} = \frac{(a \times b)(b \times c)}{(a' \times b')(b' \times c')}$$

$$\left[b'c' = \frac{bc}{\cos\theta} \right] \qquad = \cos\theta$$

$$\therefore E = \frac{I}{d^2} \cos\theta \ \text{lm/m}^2 \text{ or lx.}$$

This expression is often referred to as the cosine law of illumination.

Consider the following example.

Example

A lamp of intensity 300 candelas in all directions is placed 4 m vertically above the centre of a horizontal table measuring 3m × 3m. Calculate the maximum intensity of illumination on the table and the ratio of the maximum to the minimum illumination.

Maximum intensity at centre of the table (Fig. 10.6),

$$E = \frac{I}{(OA)^2} = \frac{300}{16} = 18\cdot8 \ \text{lm/m}^2 \text{ or lux.}$$

Minimum intensity at the corners,

$$AB = \sqrt{\frac{9}{2}} = \sqrt{4\cdot5} = 2\cdot12 \text{ m}$$

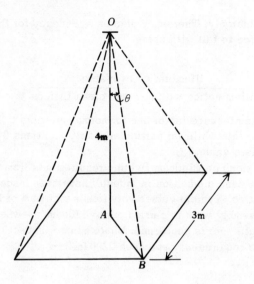

Fig. 10.6. Example.

$$E = \frac{I}{d^2} \cos \theta = \frac{I}{(OB)^2} \cos \theta$$

$$E = \frac{300}{(4^2 + 2 \cdot 12^2)} \times \frac{4}{\sqrt{4^2 + 2 \cdot 12^2}} = \frac{1200}{(20 \cdot 4)(4 \cdot 52)} = 13 \text{ lux}$$

Illumination, ratio maximum/minimum = $18 \cdot 8/13 = 1 \cdot 445$.

The whole of the light output (lumens) of a source may not be usefully used. If a lamp is suspended above a table only part of the output will fall directly on the table. Some of the light output will be absorbed by the ceiling and perhaps by the walls of the room. If a reflector is fitted this will direct light on to the table and away from the ceiling.

The ratio

$$\frac{\text{useful lumens}}{\text{total lumens emitted}}$$

is termed the utilisation factor and this may extend from less than $0 \cdot 1$ where we have indirect lighting to more than $0 \cdot 6$ for a standard dispersive open-type reflector. Tables have been compiled from which utilisation factors for specified conditions may be determined.

The illumination of a surface will also depend on whether the lighting fittings are clean — a layer of dust reduces light output or reflection considerably. There may also be a tendency for the light output of a source to fall with aging.

The ratio

$$\frac{\text{illumination available}}{\text{illumination with new and clean fittings}}$$

is termed the maintenance factor and for normal interiors 0·8 would be a reasonable for this while for particularly dirty locations 0·4 may be a more appropriate value.

Recommendations on minimum lighting requirements from the point of view of safety and production in industry have been made by various authorities. Typical values vary from a minimum of 80 lm/m² where large assembly work is carried out to 1100 lm/m² where jewellery and watch manufacture is concerned. Where minute inspection work is involved values required may approach 2200 lm/m².

Summary

Radiations which produce visible light, cover only a very small part of the electromagnetic spectrum range.

A point source of light emits *luminous flux* (Φ).

The *illumination* (E) at any point on a surface is the luminous flux received on unit area of the surface.

The unit of illumination is the *lux*.

The unit of *luminous intensity* (I) of a source is termed the *candela* (cd).

A total flux of 4π *lumens* is emitted by a light source of intensity one candela.

The *luminance* (L) of a source is measured in candelas per square metre.

When light rays illuminate a surface,

$$E = \frac{I}{d^2}$$

where d is the distance between the source and surface.

If the surface is tilted through an angle θ then,

$$E = \frac{I}{d^2} \cos \theta$$

$$\text{Utilisation factor} = \frac{\text{useful lumens emitted}}{\text{total lumens emitted}}$$

Maintenance factor = $\dfrac{\text{illumination available}}{\text{illumination with new and clean fittings}}$

Examples 10.1.

1. How does the illumination of a surface depend on the distance of a light source from the surface? Describe an experiment to illustrate this.

2. Explain the meaning of luminous intensity and state the basic SI unit. Calculate the total luminous flux emitted by a light source of intensity 200 candelas.

3. Two lamps A and B are placed 3m apart on either side of a matt white screen. The lamps have outputs of 2000 and 8000 lumens respectively, uniformly distributed. Calculate the position of the screen so that it will receive equal illumination on each side. What is the value of that illumination?

4. Two incandescent lamps of luminous intensity 100 cd and 60 cd respectively in all directions are fixed at the ends of a photometer bench. A double sided matt white screen is placed directly between the two lamps with opposite faces normal to the rays from the lamps. The face opposite the 60 cd lamp receives an illumination of 25 lux and the other receives an illumination of 62 lux. Calculate (a) the distance in metres of each lamp from the screen and (b) the illumination of each side of the screen when it is placed half way between the lamps.

5. In what units is the illumination received by a surfaced measured?

A lamp having a uniform intensity of 300 cd is suspended 3 m above the centre of a table. Determine the illumination at the centre of the table. If the lamp is raised to 4 m by how much would the illumination at the centre of the table decrease?

6. Explain the difference between luminous intensity and luminance. How would you expect the values of these quantities to compare for 100 watt tungsten filament clear and pearl lamps?

7. State the cosine law for determining the illumination of a surface inclined at an angle θ to the incident light rays.

Calculate the illumination at the corners of a table which measures 2 m × 3 m if a lamp of 250 cd uniform luminous intensity is suspended 2·5 m above the centre of the table.

8. Define the 'lumen'.

Two lamps of 500 and 300 cd are placed at points 30 m apart and 8 m above the centre of a road. Calculate the illumination at a point between the lamps 10 m from a point vertically below the 500 cd lamp.

9. A corridor is illuminated by three lamps each 3 m above the floor and spaced 5 m apart in a straight line. If each lamp has a uniform luminous intensity of 150 cd calculate the illumination on the floor (a) immediately beneath the centre lamp and (b) midway between adjacent lamps. Neglect reflection from walls and ceiling.

10. State the two fundamental laws of illumination.

A lamp having a uniform luminous intensity of 300 cd in the lower hemisphere is placed 6 m above a horizontal surface. Calculate the illumination on the surface at a point directly under the lamp and also the horizontal distance from this point at which the illumination on the surface is half this value.

11. A hall 20 m × 40 m is illuminated by 32 lamps each of 224 candelas uniformly distributed. What is the average illumination over the hall assuming a utilisation factor of 0·48 and a maintenance factor of 0·8 for the reflectors and fittings.

Filament lamps

In general, the higher the temperature of the tungsten filament in this type of lamp, the greater is the efficiency. A limit is set on the temperature however because of filament evaporation. This evaporation results in a reduction of the filament size and darkening of the glass bulb with consequent loss of light output.

If an inert gas such as argon or nitrogen is introduced, the lamp filament operates under pressure and this results in a reduced rate of evaporation compared with a similar filament operating in a vacuum. The introduction of gas however conducts heat from the filament to the glass, and hence the same power input tends to produce a lower filament temperature. This may be avoided by coiling the filament into a helix. Gas is trapped in the helix and conduction of heat takes place only from the outer surface of the helix.

The cooling area may be further decreased by using a coiled-coil type of filament. The increase in resulting temperature would lead to a shorter lamp life and to avoid this the filament length is increased slightly. A reduced power input is now required to produce the same temperature, and to maintain a standard wattage for the lamp the filament is made slightly thicker. The coiled-coil lamp filament has a greater mass than the straight and single coil filaments.

For the same filament temperature it produces a greater light output.

Lamps may be obtained with internally frosted bulbs to reduce the glare which results from the very high luminance of the light source.

An average efficiency for a 100 watt coiled-coil lamp is about 12·6 lumens per watt.

Electric discharge lamps

Filament lamps give out light as a result of the filament becoming incandescent when raised to a high temperature by the passage of an electric current. In discharge lamps the current passes through a gas which becomes ionized. Light may be produced directly by the discharge as in the sodium-vapour lamp. Alternatively, if the radiation produced is not visible to the human eye, then the radiation may be used to excite a fluorescent material which glows and emits visible radiations. Let us consider the circuits and operation of some types of discharge lamps.

Low-pressure mercury-vapour discharge lamp

Here is an experiment involving the connections to lamps of this type.

Experiment 10(iii) *To investigate the circuits of low-pressure mercury-vapour fluorescent lamps*

A lamp circuit was connected as shown in Fig. 10.7.

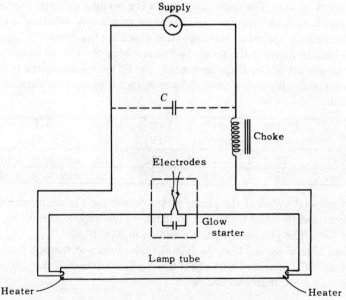

Fig. 10.7. Glow starter fluorescent lamp circuit.

The tube was of nominal length 1·5 metre and rated at 80 W. The starter unit includes two electrodes sealed into a glass bulb containing helium. A radio-active coating may be applied to the interior surface of the sealed starter to initiate the glow discharge which is necessary before the tube will strike.

When the lamp is first connected to the supply the full supply voltage appears across the starter switch. There is a discharge between the electrodes, and the heat produced causes one of the electrodes which is in the form of a bi-metal strip to bend and make contact with the second electrode. The two filaments of the lighting tube are now connected in series with the supply and a current flows. With the electrodes of the starter switch short circuited there is no longer a discharge in the starter, the bimetal strip cools and the starter contacts separate to break the supply circuit. The sudden fall in the supply current is accompanied by an e.m.f. of up to 1000 volts being induced across the choke and this high voltage across the tube causes a small amount of argon gas which is in the tube to become ionised. The mercury within the tube is vapourized, a tube discharge current flows, and with the resulting reactive voltage drop in the choke the p.d. across the tube falls to about 100 − 110 V. This voltage is sufficiently high to maintain the discharge within the tube but not to operate the starter unit, and the starter electrodes remain open while the lamp is in use. The capacitor across the terminals of the starter unit supresses high frequency oscillations which may interfere with communications circuits. The capacitor C across the complete circuit is provided to improve the power factor at which the circuit operates.

In the circuit of the lamp connected, the following tabulated results were obtained with and without the power factor correction capacitor in circuit.

	Power (W)	Supply current	Supply (V)	Lamp (V)	P.F.
With C	76	$I_1 = 0·34$ A	232	120	$\cos \phi_1 = 0·57$
Without C	76	$I_2 = 0·62$ A	232	120	$\cos \phi_2 = 0·53$

The nominal value of the power factor correction capacitor was 6 microfads and this has the effect of reducing the supply current by almost 50%. The phasor diagram is shown in Fig. 10.8.

It was also observed that the initial rush of current through the circuit with C connected was about 0·6 A while without C the initial current was more than one ampere.

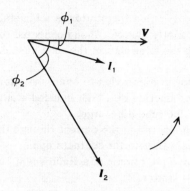

Fig. 10.8. Fluorescent lamp current with and without power factor correction.

The light output of a gas discharge lamp flickers at twice the supply frequency and this may produce undesirable stroboscopic effects when fluorescent lamps are used to illuminate rotating machines. This effect may be reduced by using lamps in pairs. One lamp is connected in series with a choke and the other with a capacitor so that the flicker effects of the two lamps are staggered.

The luminous efficiency of this type of lamp is high – in the region of 55 lumens per watt.

Here is an alternative method of starting.

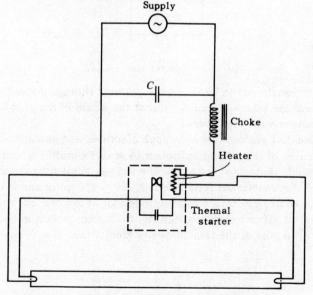

Fig. 10.9. Thermal starter fluorescent lamp circuit.

361

One of the electrodes within the starter is a bimetal strip and the two electrodes are normally closed. When connected to the supply, the initial current through the filaments of the tube passes through a small heater in the starter.

This causes the starter electrodes to open and interrupt the current with the subsequent production of a high induced e.m.f. across the choke and this starts the tube discharge.

While the lamp is in operation the current through the starter heater is sufficient to maintain the starter contacts open.

In Fig. 10.10 we have the circuit of a fluorescent lamp which does not require a separate starter.

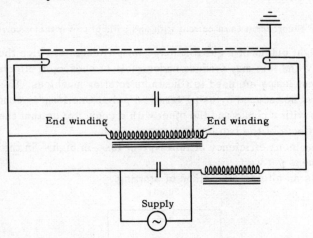

Fig. 10.10. Instant start circuit.

An auto transformer is used and the currents through the end windings preheat the tube electrodes. Almost the whole of the supply appears between the electrodes.

The potential gradient between each electrode and an earthed strip along the side of the tube is sufficient to start ionisation which eventually spreads throughout the tube. The voltage across the centre section of the transformer falls to about 100 − 110 volts and that across the heater electrodes to about one half of the original value.

The circuit gives immediate starting. No starter switch is necessary but the initial cost of the transformer is greater than that of the starter.

High-pressure mercury-vapour discharge lamp

Experiment 10 (iv) *To investigate the circuits of a high-pressure mercury-vapour discharge lamp*

These lamps are manufactured in sizes of from 50 W and an outline diagram of an 80 W unit is shown in Fig. 10.11.

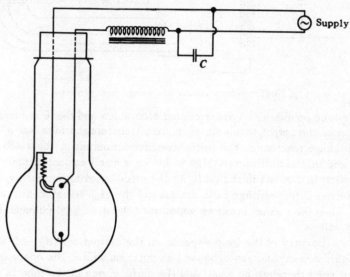

Fig. 10.11. High-pressure mercury-vapour lamp.

The lamp had two main electrodes and also an auxiliary electrode in a sealed tube containing a small amount of mercury and also argon gas. The initial discharge is through the argon gas between the auxiliary electrode and one main electrode. As the mercury becomes ionised the discharge then takes place between the main electrodes. This type of lamp may take up to 5 minutes to reach full brilliance and has a high luminous efficiency of the order of 40 lm/W.

A series choke is fitted as a stabilising impedance and a power factor correction capacitor is desirable.

The lower rated lamps of this type may be provided with a tungsten filament which in addition to replacing the series choke provides colour modification, but the efficiency of the lamp is reduced.

Low-pressure sodium-vapour discharge lamp

Experiment 10 (v) *To investigate the circuit of a sodium-vapour discharge lamp*

The diagram Fig. 10.12 shows the connections to this type of lamp.

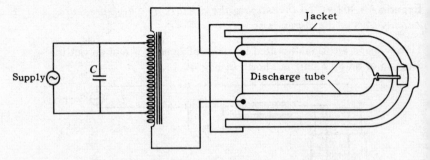

Fig. 10.12. Sodium-vapour discharge lamp circuit.

The tube contains two oxide coated electrodes and these are connected across the output terminals of an auto transformer which has a high leakage reactance. The initial transformer output is about 480 volts and this is sufficiently high to ionise a small amount of neon gas which is present in the tube. As the circuit current increases, the transformer output voltage falls because of the high transformer reactance. When the sodium becomes vapourised the discharge changes from red to yellow.

The efficiency of the lamp depends on the current density and falls off if this density increases beyond an optimum value. The operating current must therefore be small and the surface area of the tube is large. To conserve heat the tube is mounted in a double walled detachable jacket. The lamp is used horizontally to avoid the sodium collecting at one end when the lamp cools.

The efficiency of the lamp at 65 − 70 lm/W is very high. In view of the high leakage reactance of the transformer, power factor correction is important.

Light sensitive cells

Light is a form of energy and photocells have been developed which enable us to link light energy with electrical energy.

Experiment 10 (vi) *To investigate light sensitive cells*

(a) *Self generative or barrier layer cell*

The cell as examined consisted of a steel base plate with a layer of selenium covered by a thin film of transparent metal (Fig. 10.13).

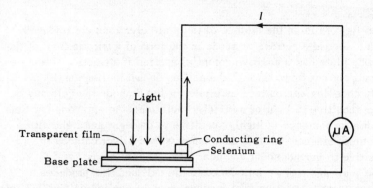

Fig. 10.13. Selenium photo-voltaic cell.

When light passes through the transparent film electrons are liberated from the selenium and accumulate on a conducting ring which makes contact with the transparent film. The potential produced may be utilised to cause a microammeter pointer to deflect. This is a true conversion of light energy into electrical energy.

High resistance units of several megohms comprising several cells connected in series have been assembled, and these may be looked upon as voltage generators. Comparatively lower resistance units of only a few thousand ohms may be looked upon as current generators. These cells have a long life and their simplicity of action makes them suitable for use in portable indicating light meters calibrated directly in terms of lux (lm/m^2).

(b) *A photo-emmisive cell*

The cell investigated consisted of a cathode plate electrode of specially prepared metal in an evacuated glass bulb (Fig. 10.14).

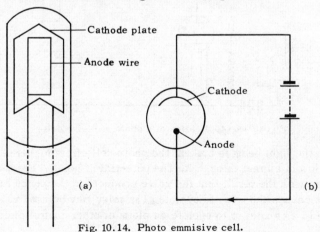

Fig. 10.14. Photo emmisive cell.

As light falls on the surface of the plate electrons are released. When a second electrode or anode in the form of a wire in front of the cathode plate has a positive potential applied it attracts electrons and a small current flows through the cell as shown in diagram (b).

By choosing appropriate materials for the cathode the cell may be made sensitive to light of particular radiations. An antimony-caesium coating for instance is highly sensitive to daylight and light with a blue predominance, while a caesium oxydised silver cathode is sensitive to incandescent and near infra-red radiations.

An inert gas may be introduced to the cell and this produces a greater current variation over a wider range of applied voltage but this type of cell is less responsive to rapid changes in light intensity than a vacuum cell.

Photo-emissive cells have many practical applications including their use in automatic counting devices, automatic opening of doors, acoustic reproduction from sound film, and in burglar alarm systems. In the latter application an invisible infra-red beam may be reflected by mirrors to cover a considerable area and any breaking of the beam causes an alarm to sound.

Figure 10.15 shows the cell connected in the grid circuit of a gas filled triode.

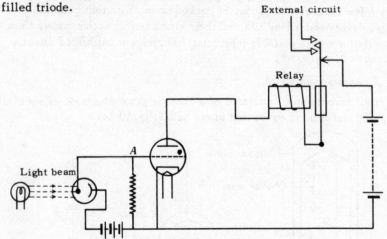

Fig. 10.15. Application of a photo emissive cell.

When the light beam is broken, the photo cell offers higher resistance to the flow of current through it. The potential of point A becomes less negative, with the result that the valve conducts in the anode circuit and this causes the relay to operate. The relay may be used to send an impulse to a counter or to operate an alarm or other control circuit.

Summary

In filament lamps, light is produced by raising the temperature of a filament until it becomes incandescent. Discharge lighting depends on a gas becoming ionised. The gas then emits either visible radiations, or non-visible radiations which may be used to act on fluorescent materials and produce visible light rays.

Low-pressure mercury-vapour fluorescent lamps commonly depend on a high voltage to initiate the discharge. In the high-pressure lamp the initial discharge is through argon gas.

In the low-pressure sodium-vapour lamp which has a high efficiency, the discharge is initiated with the aid of an auto transformer.

Discharge lamps require external reactance to limit the current when they are conducting and power factor correction is also important.

Light sensitive cells of the photo-voltaic type produce an e.m.f. when light falls on them. In the photo-emissive type of cell resistance changes when the illumination is varied.

Examples 10.2.

1. How does an electric discharge lamp differ from a filament lamp?

In a low pressure mercury-vapour fluorescent lighting circuit the tube is rated at 80 watts and the current is 0·8 ampere. In series with the tube there is a ballast choke having negligible resistance. If the supply voltage is 220 volts at 50 Hz determine the inductance of the choke, the reactive VA and the power factor of the complete circuit. Illustrate your answer with a phasor diagram.

2. With the aid of a circuit diagram describe the principle and operation of a simple fluorescent lamp circuit. Describe briefly two methods of starting.

3. Describe the construction of a high-pressure mercury-vapour discharge lamp. Use a diagram to illustrate your answer. Give a typical circuit to show how such a lamp would be connected to a supply and give details of any precautions or safeguards to be observed.

4. Describe, including a circuit diagram a sodium discharge lighting unit and explain clearly the mode of operation of the lamp and control circuit. State any advantages or disadvantages of this particular form of lighting.

5. Describe one type of illumination photometer suitable for measuring the illumination received on the surface of a drawing office table,

A lamp giving 600 cd uniformly in all directions below the horizontal is suspended 2 m above the centre of a drawing office table which measures 2m × 1m. Calculate the maximum and minimum illumination on the surface of the table assuming it is level.

11. Producing e.m.f.'s by Chemical Action; Electrolytic Deposition; Secondary Cells

Introduction

In our experiments we have used single cells and batteries to produce direct currents in electric circuits. After learning about Faraday's experiments and his discoveries of electromagnetic induction we were able to appreciate how mechanical energy could be converted to electrical energy. In the preceding chapter we have seen how an electric current is produced when light falls on a photo emissive cell and we now come to consider the production of e.m.f.'s by chemical means.

In 1789 Luigi Galvani an Italian professor of anatomy accidentally discovered that when frogs' legs hanging on copper wires came into contact with iron supports the legs twitched. After many years of investigation Alessandro Volta a physicist was able to explain this phenomenon as being caused by a current of electricity flowing in the nerves of the legs. The current was produced by an e.m.f. generated through chemical action between the two dissimilar metals iron and copper, in the presence of a salt solution in the animal tissue. Galvani had discovered the primary cell and Volta explained the action.

Primary cells

A primary cell consists of two different electrodes immersed in a conducting solution. Let us carry out an experiment with such a cell.

Experiment 11(i) *To construct and test a simple primary cell*

In this experiment plates of copper and zinc were placed in a vessel containing dilute sulphuric acid (Fig.11.1) With the switch S open and a voltmeter connected across the terminals of the cell the p.d. was observed to be 1·0 volt. When the switch was closed, to connect a resistor of 200 ohms across the terminals of the cell, it was observed that the p.d. across the load resistor decreased as tabulated.

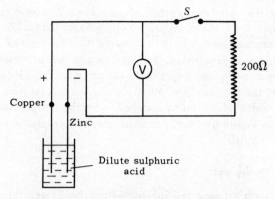

Fig.11.1. Simple primary cell.

Time (s)	0	10	20	30	40	50
P.D. (V)	1·0	0·7	0·52	0·44	0·39	0·37

These results are plotted in the graph (Fig.11.2).

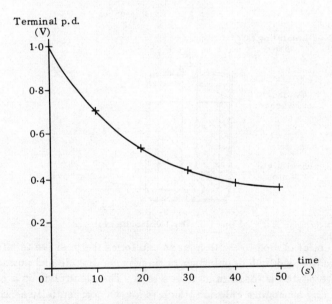

Fig.11.2. Simple cell load characteristic.

It was noticed that bubbles of hydrogen gas collected around the copper plate. The hydrogen acts as an insulator increasing the internal resistance of the cell. Also, since hydrogen is a different element from the copper it causes a separate cell to be up within the main cell. This cell generates an e.m.f. in the reverse direction to that between the copper and zinc electrodes. We say that the cell becomes polarised. The e.m.f. produced by a simple cell depends upon the electrode materials and the nature of the electrolyte. Commercial zinc has impurities present which set up local cells on the surface of the zinc plate. This is known as local action and leads to rapid decomposition of the zinc electrode.

The Leclanché dry cell

Let us examine a primary cell in which steps have been taken to avoid polarisation and local action.

Experiment 11(ii) *To examine a dry Leclanché cell*

The sketch (Fig.11.3) shows the main components of the cell examined.

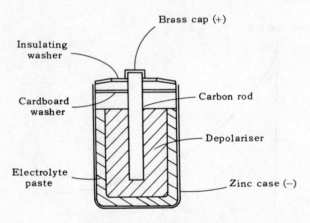

Fig. 11.3. Dry Leclanché cell.

A central carbon rod with a brass cap forms the positive terminal. It is surrounded by a depolariser of manganese dioxide and powdered graphite paste enclosed in a canvas sack. The electrolyte is a paste containing ammonium chloride which is known commercially as sal-ammoniac. An outer zinc cannister serves as the negative electrode. The e.m.f. measured with a high resistance voltmeter was found to be 1·45 volts.

To reduce the effects of local action a soluble mercury salt is added to the electrolyte. This has the effect of amalgamating or coating the zinc with mercury and rendering it less liable to local action.

The output available from the cell is reduced, depending upon the time the cell has been stored. During its 'shelf-life' the depolariser and electrolyte dry-out reducing the output available. In use the cell tends to become polarised if the current is high, but if it is only used intermittently then it recovers from this polarisation during periods of rest.

Recent developments of the Leclanché cell include the layer type of battery which is composed of successive layers of zinc, absorbent material impregnated with the electrolyte and carbon mixed with resin and a depolarising agent. Any desired voltage may be built up in this manner.

Electrolytic deposition

We have seen in a previous experiment how when plates of copper and zinc are immersed in dilute sulphuric acid an e.m.f. exists between the plates. When this simple cell is used to produce a current through an external resistor, current also flows within the cell from the zinc to the copper and hydrogen bubbles appear at the copper plate. The hydrogen is liberated from the sulphuric acid electrolyte by the passage of an electric current through it.

Now, let us carry out an experiment which was carried out by Michael Faraday.

Experiment 11(iii) *To investigate the effect of passing a current between two copper plates immersed in copper sulphate solution*

A circuit was connected as in Fig.11.4. The cell comprising the two plates immersed in an electrolyte and carrying current is known as a voltameter.

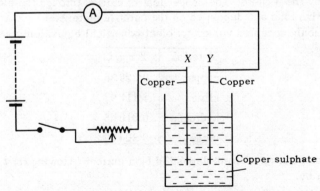

Fig.11.4. Determining electrochemical deposition.

The X plate was carefully cleaned and weighed and a current of 1·0A passed through the cell for 1000 seconds. It was observed that a deposit of copper appeared on plate X which was dried and reweighed. It was then replaced in the cell and a similar current allowed to flow for a further 1000 seconds when the plate was again dried and weighed. The observations were as tabulated.

Time (s)	0	1000	2000
Plate X (g)	31·73	32·05	32·38
Current (A)	1·0	1·0	1·0

Copper deposited during 1000 s = 32·05 − 31·73 = 0·32 g

Copper deposited during 2000 s = 32·38 − 31·73 = 0·65 g

It is thus seen that the amount of copper deposited on the X plate, or cathode is proportional to the time for which the current flows, and hence to the quantity of electricity passing through the cell.

Total copper liberated = 0·65 g

Quantity of electricity = 2000 ampere seconds or coulombs.

The mass of an element liberated when unit quantity of electricity flows through the cell is termed the electrochemical equivalent of the element.

Hence the electrochemical equivalent (E.C.E.) for copper,

$$Z = \frac{0·65}{2000} = 0·000325 \text{ g/C}$$

If the Y plate or anode is weighed before and after the experiment it will be found that this plate loses the same amount of copper as that gained by the X plate. The copper lost is carried through the electrolyte by positive ions and deposited on the cathode plate X.

Commonly accepted values for electrochemical equivalents are:

Material	Z mg/C
copper	0·329 38
silver	1·117 93
hydrogen	0·010 45
nickel	0·304 09

The mass of an element deposited by a current I flowing for t seconds is given by,

$$\text{mass} = Z \times I \times t \text{ grammes.}$$

Example

How long will it take to deposit a layer of nickel 1·0 mm thick on the cylindrical surface of a steel rod 0·25 m long and 0·025 m diameter if a current of 40 amperes is available. Density of nickel; $8·9 \times 10^3 \text{ kg/m}^3$.

Volume of nickel = area of surface × thickness

$$= (\pi \times 0·025 \times 0·25) \times \frac{1·0}{10^3} \text{ m}^3$$

mass = volume × density

$$= \frac{0·01965}{10^3} \times (8·9 \times 10^3) \text{ kg}$$

$$= 0·175 \text{ kg or } 175 \times 10^3 \text{ mg}$$

Mass $= Z \times I \times t$

$$I \times t = \frac{175 \times 10^3}{0·304} \text{ C}$$

$$t = \frac{175 \times 10^3}{0·304 \times 40} \text{ s} = 14400 \text{ s}$$

Time = 4 hours

Applications of electrolysis include electroplating and the refining of metals. In electroplating the unit on which the deposit is to be made is connected as the cathode and the element to be deposited is the anode. Both are immersed in a suitable electrolyte and a current passed through the cell.

In refining, the impure metal is the anode and the refined metal is depositied at the cathode. Impurities fall to the bottom of the cell as sludge.

Summary

A simple primary cell consists of two dissimilar metals immersed in an electrolyte.

In the Leclanché primary cell manganese dioxide is the depolarising agent and a deposit of mercury on the zinc reduces local action.

The electrochemical equivalent of an element is the mass of the element liberated by unit quantity of electricity.

$$\text{mass} = Z \times I \times t$$

Examples 11.1.

1. Describe the construction of a Leclanché cell.
   Why is a depolariser used in this cell? What value of e.m.f. can be expected from a single Leclanché cell when the cell is new and when it is subjected to loading over a long period?

2. What is meant by the electrochemical equivalent of an element? Describe an experiment which may be carried out in the laboratory to determine the electrochemical equivalent of copper.

3. A 12 ohm resistor is connected in series with a copper voltameter and the current in the circuit is maintained at 5 A for 10 minutes and then at 2 A for 15 minutes. If the electrochemical equivalent of copper is 0·32 mg/C calculate the mass of copper deposited and the energy dissipated in the resistor. What steady constant value of current flowing for 45 minutes would deposit the same amount of copper?

4. How does the conduction of electricity through an electrolyte differ from conduction through a copper wire?
   In an experiment a current of 1 ampere was passed between two electrodes immersed in a solution of silver nitrate and 6·72 grams of silver were deposited in 100 minutes. Calculate the electrochemical equivalent of silver.

5. When a current of 1 ampere is passed through an electrolytic cell for 10 minutes 0·67 g of metal is deposited at the cathode. Determine the amount deposited when a current of 2 amperes is passed for 5 minutes and also the electrochemical equivalent of the metal.

6. Copper is purified in a vat which takes a current of 1200 amperes at a terminal p.d. of 4·8 volts. Determine the mass of copper deposited per hour and the electrical energy required per kilogramme of copper deposited.

7. Determine the time required to deposit a thickness of 0·5 mm of nickel on the surface of a worn steel shaft which has a diameter of 0·2 m and a length of 4 m. The current density is to be maintained at 200 A/m². Assume an efficiency of 95%. Density of nickel = $8·9 \times 10^3$ kg/m³.

8. Describe with the aid of a sketch one process of electrolytic deposition of a metal. Show clearly the direction of current flow.
   A copper plating bath is used to check the accuracy of an ammeter. The ammeter indicates 3 A and 1·14 grammes of copper are deposited in 20 minutes. If the electrochemical equivalent of copper is 0·329 mg/C determine the ammeter error.

Secondary cells

The commonest primary cell in everyday use is the Leclanché dry cell. Many other primary cells have been developed with alkaline, acid and water activated electrolytes. All primary cells have the common property that the chemical reactions which occur during discharge are not reversible. In secondary cells chemical reactions which take place during discharge may be reversed by passing a charging current through the cell, in the reverse direction to the discharge current. The internal resistance of secondary cells is lower than that of primary cells.

The lead-acid cell

The action of a lead acid cell is illustrated by the diagram in Fig.11.5.

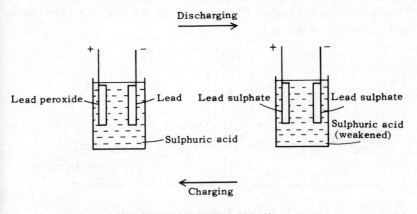

Fig.11.5. Lead-acid cell.

When fully charged the plates consist of lead peroxide and lead. They are immersed in sulphuric acid which serves as the electrolyte. During discharge the plates absorb sulphate ions from the acid and are converted to lead sulphate. Hydrogen released from the acid combines with the oxygen from the positive plate to form water. The resulting electrolyte in the discharged cell is less concentrated than when the cell is charged.

During charging, a current is passed in the reverse direction through the cell and a process of electrolysis results in the electrodes being converted once again into lead peroxide and lead.

The physical construction of the cell or battery depends upon the purpose for which it is intended. In portable cells as used in vehicles, it is essential that the battery has a high ratio of electrical capacity to weight. When cells are stationary however and used for emergency lighting, or switch tripping operations in power stations, durability is of prime importance and weight is a secondary consideration.

The Fauré or pasted construction consists of a lead grid into which the active material is pressed. A comparatively light plate results. In the Planté or formed construction lead plates are repeatedly subjected to charging and discharging until eventually active material is accumulated. The process may be accelerated using a catalyst.

Here is an experiment to plot the charge and discharge characteristics of a lead-acid cell.

Experiment 11 (iv) *To plot the charge and discharge characteristics for a lead-acid cell*

The cell to be tested was fully charged and with a rated capacity of 55 ampere hours it was then discharged at the ten-hour rate. The circuit was connected as in Fig.11.6.

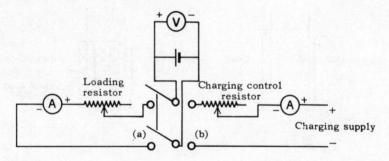

Fig.11.6. Load testing a lead-acid cell.

With the two way switch in position (a) the loading resistor was adjusted to maintain the load current at $55\,Ah/10\,h = 5\cdot5\,A$ throughout the period of discharge.

The instrument readings were as tabulated.

Time (h)	0	1	2	3	4	5	6	7	8	9	10
I (A)	5·5	5·5	5·5	5·5	5·5	5·5	5·5	5·5	5·5	5·5	5·5
P.D. (V)	2·01	1·99	1·98	1·97	1·96	1·94	1·93	1·91	1·89	1·87	1·84

These results are plotted in Fig.11.7.

With the two way switch in position (b) the cell was then recharged with a constant charging current of 5·5 A. The charging voltage was observed at intervals and tabulated. After about 11 hours the cell was gassing freely and the p.d. across the cell terminals settled at 2·74 volts

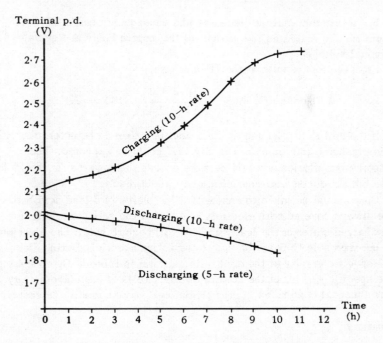

Fig. 11.7. Lead-acid cell characteristics.

This free gassing and steady p.d. across the terminals indicated that the cell was fully charged.

Time (h)	1	2	3	4	5	6	7	8	9	10	11
I (A)	5·5	5·5	5·5	5·5	5·5	5·5	5·5	5·5	5·5	5·5	5·5
P.D. (V)	2·16	2·18	2·2	2·26	2·32	2·4	2·5	2·6	2·7	2·73	2·74

The charging curve is also plotted in Fig. 11.7.

Efficiency

The efficiency of a battery may be given in terms of ampere-hours or watt-hours.

$$\text{Ampere-hour efficiency} = \frac{\text{ampere hours available during discharge}}{\text{ampere hours to replace the charge}} \times 100\%$$

From the test results at the 10-hour rate,

$$\text{ampere-hour efficiency} = \frac{10 \text{ Ah}}{11 \text{ Ah}} \times 100 = 91\%$$

$$\text{Watt-hour efficiency} = \frac{\text{watt hours available during discharge}}{\text{watt hours to replace charge}} \times 100\%$$

For a constant-current discharge and subsequent charge the watt hours may be stated as the product of the ampere hours and average terminal voltage.

From the test results at the 10-hour rate,

$$\text{Watt-hour efficiency} = \frac{10 \times 1\cdot92}{11 \times 2\cdot43} \times 100 = 72\%$$

If the cell is discharged at the 5 hour rate then a characteristic with this discharge rate as shown in Fig.11.7 could be expected. The ampere-hour efficiency would be more or less the same as for the 10 hour rate but a reduced watt-hour efficiency would result.

Care should be taken to ensure that the plates of a lead-acid battery are always covered with electrolyte, and that the cell is not left in a discharged state, or the lead sulphate will become hard. The electrolyte is not then able to pass through to the active material During discharge the specific gravity of the acid falls as water is formed. By comparing the specific gravity of the acid in different cells of a storage battery any abnormal conditions in a particular cell may be readily detected.

Summary

In a secondary cell the chemical reactions which occur during discharge may be reversed by passing a charging current in the reverse direction through the cell.

The efficiencies of secondary cells may be expressed in terms of ampere hours or watt hours.

Examples 11.2.

1. Describe carefully with the aid of a circuit diagram how you would proceed to determine the charge and discharge characteristics for a lead-acid cell.

Sketch typical characteristics for a 12 V lead-acid battery during charge and discharge at the 20-hour rate. How would you expect curves at the 10-hour rate to compare with those at the 20-hour rate?

2. What is meant by the ampere-hour and watt-hour efficiencies of a storage battery?

A battery of 100 cells in series takes a constant charging current of 8A and the e.m.f. rises from $1\cdot8$ volt to $2\cdot4$ volts per cell. If each cell has an internal resistance of $0\cdot006\,\Omega$ calculate the p.d. across the battery at the beginning and end of the charge.

If the battery is completely discharged in 40 hours at 4A with an average p.d. of $1\cdot95$ volts per cell, after being charged for 24 hours at

8 amperes with an average p.d. of 2·2 volts per cell, determine the ampere-hour and watt-hour efficiencies of the battery.

3. Describe the processes of charge and discharge in a lead-acid secondary cell. How would you expect the construction of the plates for cells of a large stationary battery to compare with those of a portable battery?

A battery of 80 lead acid cells each having an e.m.f. of 2·0 V and internal resistance of 0·005 Ω is connected for charging as in Fig.11.8. Determine the resistance R_1 and R_2 if the charging rate is to be 24 A for the 50-cell section and 12 A for the 30-cell section.

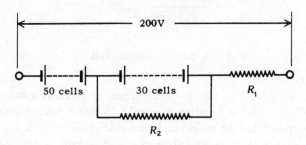

Fig.11.8. Example 3.

4. A 12 volt lead-acid accumulator has a rated capacity of 45 ampere hours at the 20-hour rate. Its terminal voltage falls from 13·2 volts to 10·8 volts during discharge. While being charged with a current of 4 amperes the terminal voltage rises from 12·6 volts to 16·2 volts. It is proposed to charge the accumulator at 4 amperes from a source having a constant terminal voltage of 24 volts. Draw circuit diagrams for the charge (at 4 A) and discharge (at the 20-hour rate) conditions and calculate the limiting values of resistance required in each circuit.

Alkaline cells

Alkaline nickel-cadmium cells have potassium hydroxide as the electrolyte nickel hydroxide as the positive plate, and cadmium as the negative plate. The outer container is of steel with welded joints and treated to prevent corrosion.

The action during charge and discharge is illustrated by the diagram (Fig.11.9.)

During discharge cadmium is converted to cadmium hydroxide and the positive nickel hydroxide $Ni(OH)_3$ changes to another form of hydroxide $Ni(OH)_2$. The electrolyte undergoes no apparent change during charge and discharge and the specific gravity remains constant at between

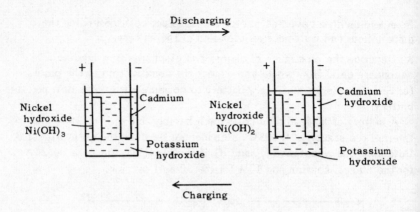

Fig. 11.9. Nickel cadmium cell.

1·16 and 1·19.

The nickel-iron cell is similar to the nickel cadmium cell but iron replaces cadmium as the negative electrode. A rather lower charging voltage is required for the nickel-cadmium cell since it has a lower internal resistance than the nickel-iron cell. It is used in applications where the difference between the charge and discharge voltages is important and where a high rate of discharge is required at a comparatively well maintained voltage.

Here is an experiment on a nickel-cadmium cell.

Experiment 11(v) *To determine the discharge characteristic for a nickel cadmium cell*

A similar circuit to that used in **Experiment 11(iv)** was connected. The cell tested was a nickel cadmium cell of 50 Ah capacity and it was discharged at the 5 hour rate. Observations of the p.d. across the cell terminals were tabulated and are plotted in Fig. 11.10.

Time (h)	0	1	2	3	4	5
I (A)	10	10	10	10	10	10
P.D. (V)	1·25	1·23	1·22	1·2	1·19	1·12

Comparison of lead-acid and alkaline cells

Lead-acid cells operate with rather higher ampere-hour and watt-hour efficiencies than alkaline cells. This is partially because of the increased gassing which takes place during the charging of alkaline cells. The capacity of the alkaline cell does not decrease on high

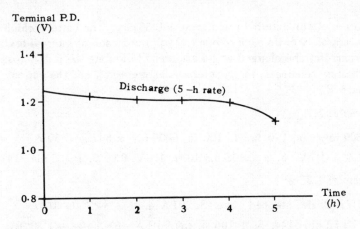

Fig. 11.10. Ni–Cd cell discharge characteristic.

rates of discharge to the same extent as that of the lead-acid cell.

To obtain equal output voltages from batteries on discharge, more alkaline cells than lead-acid cells are necessary. The initial cost of an alkaline storage battery is higher than that of the lead-acid battery of similar capacity. Alkaline cells have a longer life and there is no sulphation of the plates. Under heavy discharge currents the plates will not buckle and their construction is more resistant to vibration than the plates in lead-acid cells.

Examples 11.3.

1. Describe with the aid of diagrams the action of a nickel-cadmium cell during charge and discharge. How do nickel-iron and nickel-cadmium cells differ?

2. How would you expect the ampere-hour efficiencies of lead-acid and alkaline cells to compare?

Twenty-four alkaline secondary cells, each of discharge capacity 80 ampere hours at the 10-hour rate are to be charged with a constant current for 8 hours. Calculate the maximum and minimum values of the charging resistance required if a d.c. supply of 50 volts is available and the ampere hour efficiency of the cells is 80%. The e.m.f. of each cell at the commencement and end of charge is 1·3 volts and 1·7 volts respectively.

3. Compare a lead-acid cell and an alkaline cell with regard to; component parts, charge and discharge characteristics, weight, durability and life, fields of use and maintenance.

4. A separately excited generator charges a battery and supplies current to a 10 ohm resistor R in parallel with the battery. The generator

armature circuit internal resistance is 0·55 ohm. The battery which consists of 10 cells each of e.m.f. 2 volts and each of internal resistance 0·06 ohm is to be charged at 2·5 amperes. Calculate the p.d. across the battery terminals, the generator armature e.m.f. and the current through R.

Examples 1.1.

2. 500 rev/min, 120 Hz. 4. 100 Hz, 6000 rev/min, $E_m = 60$ V.
5. $E_m = 1·7$ V. 6. $e = 196 \sin 105 t$, 170 V, 0 V. 7. $i = 6 \sin 314 t$.

Examples 1.2.

2. 1·23, 25 Hz, stepped wave. 3. 7·1 A 4. $\nu = 84·7 \sin 314 t$,
$i = 14·12 \sin 314 t$. 5. 1·106 6. (a) 70·1 A, (b) 50 Hz, (c) 30·9 A.
7. 1·16, 1·73

Examples 1.3.

1. −20·2 V, 594°. 2. V_2 leads 3. 13·4 sin $(\theta + 0·25)$, 0·25 rad, 0·0008 s.
4. 81 sin $(\omega t + 0·12)$. 5. 79·4 sin $(\omega t + 0·88)$. 6. 96·5 sin $(\omega t - 0·175)$.

Examples 2.1.

1. 576 W. 2. 96·8 Ω (a) 1000 W, (b) 0 W.

Examples 2.2.

2. 100 sin 628 t. 3. 10 Hz, $I = 7·94$ A; 100 Hz, $I = 0·794$ A

Examples 2.3.

1. $V_R = 62·4$, $V_L = 78·4$. 2. 2540 Ω. 3. 100 Hz, 250 V, 177 V, p.f. 0·5
4. 12·5 Ω, 50 Ω. 5. 0·85 6. 0·76, 78·5%, 37 Ω. 7. 12·5 Ω. 8. 1·2 A,
1·11 A. 9. 5·85 kW, 35 A. 11. 142 V 12. 50 Ω, 0·5 W, 100 Ω. 13. 0·77,
6050 V, 640 V. 14. 6000 A/s, 420 V, 296 V. 15. 1·69 A, 0·237, 91·5 V,
160 V. 16. 7·7 Ω, 0·0445 H, 16 Ω, 0·48 17. 1·08 A.

Examples 2.4.

1. $I_m = 3·14$ A. 2. 10 Hz, $I_m = 0·0314$ A; 1000 Hz, $I_m = 3·14$ A
4. 31·8 μF.

Examples 2.5.

1. $V_C = 145·5$ V, $V_R = 137·5$ V. 2. 8·85 Ω. 3. 564 Ω 0·53, 20 V r.m.s., 28·4 V max.

Examples 2.6.

1. 2·04 A, V_R = 204 V, V_L = 192 V, V_C = 65 V. 2. 3·75 A, 359 V, 120 V. 3. 127 V, p.f. 1·0 4. 2·03 pF, 0·008 W, 628 V. 5. 250 Ω, 1·59 H, 6·38 μF. 6. 5·75 kHz, 1260 Hz, 0·1 A 7. 16·8 Hz, V_R = 280 V, V_L = V_C = 1265 V. 8. 63·5 μF, V_R = 110 V, V_L = V_C = 550 V, leading p.f. 9. 755 V. 10. 44·2, 112·5 Hz.

Examples 2.7.

1. 0–550 Ω. 2. 48 A, 53°; 26·8 A, 26°; 17·7 A, 17°. 3. 1·25 A; 0·707, 0·885 A. 4. 7·2 A, 4·6 A, 1570 W.

Examples 3.1.

1. I_R = 4·8 A, I_C = 3·02 A, I = 5·66 A, 1150 W, 0·85. 2. 19·7 μF, 0·625. 3. 2·8 A, 0·97. 4. 0·98 lag, 0·8 lead. 5. 29·8 A, 0·623, 4480 W; 13·3 A, 0·5, 1780 W; 24 A, 1·0, 5760 W; 51·0 A, 0·98 lag. 6. 16 kVA, 0·95 lag. 7. 1·08 V, 0·5 mA, 1·47 mA, 0·97 lead, 1·65 mW. 8. 5·35 A.

Examples 3.2.

1. 0·5 MHz. 2. 415 μF. 3. 49·6 Hz. 4. 5·65 V, 36 mW, 0·916, 0·05 μF.

Examples 3.3.

1. (a) 8·95 ∠63° (b) 5 ∠53° (c) 10·6 ∠229° (d) 13·4 ∠–63° (e) 10·8 ∠–22° (f) 8·25 ∠166° 2. 10·4 + j6 3. 398 μF.

Examples 3.4.

1. (a) 12·6 ∠71·5° (b) 3·2 ∠–71·5° (c) 2·8 ∠–45° (d) 25·6 ∠69·5°
2. (a) 2·83 ∠45° (b) 7·07 ∠–80° (c) 12·3 ∠189·5° (d) 14·6 ∠–16°
3. 13·4 V 63·5°

Examples 3.5.

1. 86 Ω, 45·6° inductive. 2. 0·26 A, 0·55. 3. 146 V

Examples 3.6.

2. 63·4° 26·6° 60 V ∠90° 3. 300 ∠106 4. 0·695 ∠107° 5. 16·4 ∠55·4°
6.
$$\frac{Z_1 Z_2}{Z_1 + Z_2}, \quad \frac{Z_1 Z_2}{Z_1 + Z_2} + Z_3$$

7. 44·5 Ω, $R = 425$ Ω, $X_C = 12$·55 Ω 8. (a) 51·8 W 49·1 VAr. (b) 860 W 252 VAr.

Examples 3.7.

1. 39 Ω, 0·0256 s, 0·01965 s, 0·0164 s, 0·77. 2. 0·026 s, 0·01375 s, 0·0294 s, 0·0312 s, 0·00826 s, 0·0324 s, 3·24 A. 3. 0·52 21 A, 0·748.
4. 0·135 s, 0·133 s, 7·4 Ω, 0·0257, 7·26 Ω, 1·4 Ω, 7·4 Ω. 5. 2·64 Ω, 2·5 Ω, 42·64 Ω, 12·5 Ω.

Examples 4.1.

3. 208 V, 1·2 A, 0 A. 4. Voltage = $2E_{ph}$ 5. 750 rev/min, 200 V, 115 V, 6·9 A, 200 V, 4 A, 6·9 A. 6. 5·6 A, 9·65 A, 240 V, 58°. 7. 34·4 A.

Examples 4.2.

5. 2490 W, 4·54 A, 231 V. 6. 6·86 kVA, 4·84 kW; 20·5 kVA 14·5 kW.
7. 55 W, 425 W.

Examples 4.3.

1. 33·3 Ω 2. 14·5 Ω, 0·69 3. 13·85 + j11·4, 12·4 − j3·76, 11·4 − j13·85
4. $R = 90$ Ω, $X_L = 75$ Ω. 5. 65 + j17, 30·4 + j88, 45·3 − j64·2

Examples 5.1.

1. 1800 rev/min. 3. 2·67%, 1·33 Hz. 5. 1440 rev/min, 1000 rev/min.
6. 2550 kWh, 42·7 A, 254 V. 8. 5·4 A, 9·4 A. 10. 10·1 A, 17·3 A, 440 V, 254 V. 11. 11·55 A, 3560 W, 0·291. 12. 48·2 A, 27·8 A, 440 V, 254 V.
13. 28·5 A 14. 950 rev/min.

Examples 5.3.

1. 106 μF, 35·7 A, 27·8 A. 2. 162 A, 120 A 3. 245 μF. 4. 172 μF.
5. 19·2 kW, 31·4 kVA, 131 A, 80 A, 104 A, 1385 μF 72 A ∠46°.
6. 520 μF. 7. 0·842 p £350 8. 1·65 p 9. £3620, £3500, 560,000 kWh.

Examples 6.1.

1. 40 : 1, 120 : 1. 2. 2200 V, 11·4 A. 3. 8·33 V, 0·00125 Wb. 4. (a) 5400 (b) 281 5. 0·248 T 1100 V. 6. 73, 31·25 A. 7. 0·3125 A, 1·21 A.
8. 0·12 A, 0·1965 A. 9. 169, 776. 10. (a) 0·168 H, (b) 1·34 V, (c) 0·125 H 11. 0·1565 H, 0·106 H, 12. (a) 1·25 mH, (b) 0·625 V
13. (a) 0·09 H, (b) 0·0675 H (c) 0·315 H 14. 25 Hz. 15. 244.

16. (a) 280 mH, (b) 120 mH. 17. 0·53, 28 mH.

Examples 6.2.

2. 2:1, 55·5 Ω. 3. 350, 467, 8·33 A 4. 1500 V, 115 V, 13·3 A, 174 A..
5. 11 A 6. 12·8 A, 0·768 11. 95·8%, 95%.

Examples 6.3.

2. 0·192 Ω. 3. 0·6 Ω, 0·475 Ω, 0·765 Ω 4. 938 5·62 A, 7·5 A 17·8 Ω.
5. 768 Ω, 815 Ω 6. 0·678 Ω, 0·548 Ω. 7. 94·5% 9 V 8. (a) 97·2%,
96·8% (b) 3·14%

Examples 6.4.

3. 6·4 W/kg 4. 27·3 A, 0·8 5. 2100 W, 6300 W.

Examples 6.5.

3. 17 A, 3 A.

Examples 7.1.

1. 1·53 V 2. 0·65 m, 0·74 mA 3. 0·102 Ω. 4. 1 Ω. 5. 28·4 A 7. 0·162
of length. 8. 0·008 A 9. 10·1.

Examples 7.2.

3. 14·75 A

Examples 7.3.

1. $28·8 \times 10^{-7}$ Nm 2. 120×10^{-8} Nm, 120° 3. $6·67 \times 10^{-6}$ Nm 4. (a) 2 mA,
(b) 200 kΩ, (c) 10 mV. 5. 16·662 Ω, 0·005 Ω, 6.199 kΩ in series.
7. 46·7 Ω, 42 Ω, 1·05 mW. 8. 0·46 V; 0·46 A. 9. 75×10^{-5} Ω, 13·5 A
10. 19·92 kΩ, 199·92 kΩ, 1999·92 kΩ. 11. 33·3 V, 50% 12. 3000 Ω.
13. 10 mA ammeter, 50 μA voltmeter, 0·49 Ω, 19·99 MΩ. 14. 75 Ω;
10 kΩ, 100 kΩ, 1 MΩ.

Examples 7.4.

3. (b) 7·05 A, (c) 5 A. 4. (a) 1·11 × r.m.s. value. (b) r.m.s. 5. 0·556 Ω,
0·0505, 0·00501 Ω.

Examples 7.5.

7. 1·8 MΩ.

Examples 8.2.

3. 29 400 Ω, 1·7 mA/V, 50. 4·1 mA/80 V. 5. 2·4 mA/1·5 mA, 34·8 V$_m$.
9. (a) 0·223 A (b) 0·107 A.

Examples 9.1.

5. 2·8 V, 4·6 V, 35. 6. 37·5, 32·5. 7. 14·033 dB; (a) 1·26, (b) 1·58, (c) 10.

Examples 10.1.

2. 2520 cd. 3. 2 m, 1m. 4. (a) 1·27 m, 1·55 m; (b) 50·3 lx, 30·2 lx.
5. 33·3 lx, 14·5 lx. 7. 21·4 lx. 8. 2·07 lx. 9. 21·21 lx. 16 lx. 10. 8·34 l; 4·59 m. 11. 43·1 lx.

Examples 10.2.

1. 0·78 H, 156·8 VA, 0·45. 5. 150 lx, 100 lx.

Examples 11.1.

3. 1·535 g, 223 kWs, 1·775 A. 4. 1·12 mg/C. 5. 1·12 mg/C, 0·67 g.
6. 1430 g/h, 4 kWh. 7. 21·5 h. 8. 0·11 A.

Examples 11.2.

2. 184·8 V, 244·8 V, 83·5%, 74%. 3. 1·34 Ω, 5·15 Ω. 4. 2·85 Ω, 1·95 Ω; 7·2 Ω, 5·6 Ω.

Examples 11.3.

2. 1·5 Ω, 0·74 Ω. 4. 21·5 V, 24·06 V, 2·15 A.

The author has planned a predominantly experimental and practical introduction to the basic principles of electrotechnology to interest and stimulate newcomers over a wide range of courses both at home and overseas. The subject matter is introduced as simply as possible; the author assumes no previous systematic knowledge of electrotechnology yet aims to establish a firm foundation of electrical engineering principles. The author is particularly mindful of the needs of students overseas who often study under conditions which preclude them from formal courses or in circumstances where workshop and laboratory facilities are few and inadequate.

Fundamentals of Electrotechnology 2: Practical Guide includes over 120 experiments, and actual laboratory test results are given. Whenever possible an experimental approach is used to develop in a logical sequence the main tenets of the basic theories underlying the science of electrotechnology. Worked examples have been included wherever useful and necessary and, as almost all experiments include calculations, the experiments in themselves form worked examples. Problems for students to solve appear at appropriate intervals in each chapter and answers to numerical questions are given at the end of the book.

The contents cover the essential requirements of the major British courses at technician level, and are relevant and suitable for use in the technician courses now being established by autonomous overseas examining authorities (such as the West African Examinations Council). Additionally, the book will match the needs of some students in first-year undergraduate or diploma courses.

£2·00 net
IN UK ONLY

ISBN 0 09 113441 2

Cover design by Brian Denyer

Printed in Great Britain by
The Anchor Press Ltd, Tiptree, Essex